Fundamentals of Telecom/Datacom

By: Paul A. Rosenberg

Edited by
John Paschal, P.E.

Fundamentals of Telecom/Datacom
Copyright © 2000 Intertec Publishing Corporation
All rights reserved

First Printing: May 2000

Published by
EC&M Books
Intertec Publishing Corporation
9800 Metcalf Avenue
Overland Park, KS 66212-2215

Library of Congress Catalog Card Number: 00-103644
ISBN 0-87288-762-6

Table of Contents

Chapter	Page
Introduction	vii

1. Basics — 1
What is datacom, what is telecom? — 1
A Brief History — 1
Signal Transmission — 2
Copper Cabling — 2
Optical Transmission — 3
Networking — 4
The Limits of Copper Wire — 5
Interference — 5
Two Types of Transmission Problems — 6
Pulse Spreading — 6
Categories? Levels? What are they? — 7

2. Networks — 9
Ring Topology — 11
Star Topology — 11
Bus Topology — 13
Attenuation — 13
Cable Impedance — 13
Capacitance — 14
Cross-talk — 14
Cable Shielding — 14
Common Types of Network Cabling — 15
Wireless Transmission — 16
Infrared Transmission — 16
Powerline Carier Networks — 17
Other Transmission Means — 17
Inexpensive Cables for High-speed Networks — 18
Ethernet — 19
Standards — 19
Collisions — 19
Devices — 20
Problems — 20
Frames — 21
How Ethernets Are Used — 21
Other Network Terms — 22

	Structured Cabling & Ethernet	24
	Network Architectures	25
	Common Architectures	26
	Physical Structure	27
	Other Design Issues	30
3.	**Cable Installation**	**33**
	Roughing and Trimming	33
	Wiring Layouts	34
	Cable Colors	35
	Separation From Sources of Interference	35
	Minimum Bending Radii	36
	Installation Requirements	37
	Conductors Entering Buildings	37
	Circuit Protection	37
	Interior Communications Conductors	40
	Requirements of Article 725	40
	Definitions	41
	Requirements	41
	Requirements	42
	Cabling Classifications	44
	Cable Specifications	45
	Telecommunications Outlet Specification	47
	Category 5 Cabling	48
	Standard Networking Configurations	48
	Ethernet 10BASE-T Cabling and Patch Cords	48
4.	**Optical Fiber**	**51**
	Fibers	52
	Fiber Sizing	52
	Cabling	53
	Fiber Connectors	54
	Splices	55
	Testing	55
	Other Basic Concepts	56
	Installation	57
	Saftey Precautions	58
	The Fiber Loss Budget	58
	Design Short-Cuts	58
	Fiber Choice	58
	Cable Design	59
	Fiber Performance	59
	Cable Performance	60
5.	**Testing**	**61**
	Components	61

What Testing Proves	61
How To Start	62
Understanding Decibels	63
When To Test	63
Common Cable Test Equipment	63
Testing UTP Cables	64
Wiremapping	65
Length Tests	66
Testing for Impulse Noise	67
Near-End Crosstalk (Next)	67
Attenuation	67
Testing Optical Fiber	68
Power Meter Testing	69
Visual Cable Tracers and Fault Locators	69
Microscopes	69
Continuity Testing	71
Optical Time Domain Reflectometer	71
The Limitations of OTDR Use	72
When To Use The OTDR	73

6. Outside Plant Installations — 75

Burying Cable	75
Underground Cable Installation	76
Underground Raceways	76
Blown-In Fiber	77
Routing of Outdoor Aerial Circuits	77
Messenger Cables	78
Underground Cable Locators	78
Directly Buried Optical Cables	79
Optical Aerial Cables	79
Hybrid Networks	79
Backbone Systems	79
Campus Systems	80
Types of Fiber	80
Installation	80
Types of Fiber Cables	80
Industry Cable Standards	81
Connectors and Splices	82
Breakout Kits	82
Manholes, Vaults and Poles	83
Fiber Optic Standards	83
Choosing Fibers	84
Choice of Cables	85
General Factors	87

7. Telephone Systems & Data Transmission — 89

The end of the phone company?	89
What does this mean to the electrical contractor?	90
Understanding Telephone Networks	90
Data Over the Telephone Lines	95
Modem Traffic	95
T1 Lines	95
The CSU	96
The DSU's Duty	96
Computer Telephony Terms	97
The Operation of Telephones	99
Digital Telephony	102
Understanding Analog and Digital Lines	102
Bandwidth and the 4kHz Channel	103
Dedicated Lines, Switched Lines	104
Phone Services	105
ISDN	106
ADSL Technology	108
Tech Details	110
The Importance of ADSL	110
Will this really happen?	111
Other Digital Services	112
Multiplexing	112
Analog To Digital Conversion	115

8. The Internet — 117

What is the Internet?	117
How It Began	117
Multiple Paths	118
Is the *NET* the same a the *WEB*?	118
The Importance	119
Spontaneous Generation	120
How To Use It	120
Here to Stay	120
The Internet's Flavor	121
Some Internet Terms	121
Encryption	122
Public Key Cryptography	123

9. The Business — 125

Getting Business	125
Advertising	125
Training	126
Estimating	127
The Take-Off	128
Labor Expense	129
Modifying Labor Units	130

Subcontracts	130
Charging Training	131
Overhead Percentages	131
Bidding	131
A New Company	132
Company Structure	133
Infrastructure	133
How Long It Takes	134
First Steps	134
Supervising	135
Inspection	135
Business Skills	136

Glossary of Terms   137

Introduction

Not long ago, data communications was a very minor part of the electrical industry; a few people specialized in it, and that was about all. Now, however, data communications is a very big part of the electrical industry, accounting for a large percentage of all electrical work.

This book is necessary for a number of reasons; but first among them is that there is a real need for a communications book that is accessible to people with electrical power backgrounds; that is, electricians and electrical construction professionals. That is my background, and I know from personal experience how difficult it is to make sense out of most communications texts, not to mention the manufacturers' spec sheets. I have made every effort to make this book understandable; datacom terms are defined, and the technologies explained. I teach seminars on datacom to electricians, and my experience answering the questions of these men and women has contributed to these pages.

This book is written to help the people who actually install communications systems. It does contain a fair amount of design information as well, but the primary focus is on understanding and installing these systems correctly.

From here on in the electrical industry, understanding data communications will generally mean that you are worth more to your employer and to your customers. It all comes down to fact that if you know more, you are far more likely to be able to buy a nicer home, a better education for your kids, nice vacations, better food, or simply a more comforting bank balance.

Datacom work is increasing continually, while power wiring is increasing slightly and cyclically. While power wiring still provides a good living to hundreds of thousands of people (and will continue to do so), datacom provides a good living, frequently a better living, and almost infinite opportunities for advancement. The datacom field expands, modifies, and changes on an almost daily basis. Competent people are in high demand. Retraining people for a specific job is an every-day thing. So, if you want to improve your situation, and are willing to expend some time and effort in so doing, datacom is where to do it. It's that simple.

Chapter one in this book covers most of the basic information on data communications: How these systems work, which technical details are important, and so on.

Chapter two covers various methods of transmitting signals through computer networks. It is important for anyone involved with data communications to understand what networks are, how we send signals through them, and be able to understand the basic technical terms.

Chapter three covers cable installation, chapter four optical fiber, and chapter five, the testing of cabling systems.

Chapters six explains the mechanics of installing cabling outdoors, which has special requirements.

Chapters seven and eight cover the new types of data communications: data over telephone lines, and the Internet.

Our final chapter, the ninth, covers the business side of the datacom industry. This chapter is not absolutely necessary for installers (as opposed to contractors and supervisors), but it is worth installers reading nonetheless. It is important to understand the business of which you are a part of.

The glossary we have included in this book is quite thorough, and should (hopefully) contain the definitions of all the difficult terms you may come across.

I hope that you will find data communications interesting, exciting, challenging, and profitable.

Welcome aboard.

– Paul Rosenberg

Chapter 1

What is datacom? What is telecom?

Telecom is short for telecommunications - the transmission of telephone signals from place to place.

Datacom is short for data communications - the transmission of data (computer) signals from place to place.

These terms are usually interchangeable. There is a good reason for this: Telephone systems are changing into data systems. Right now, Internet traffic over telephone lines is equal to or exceeding that of voice traffic. So, with the telephone system becoming the data system, it is very difficult to say whether anything attached to a phone line is telecom or datacom.

At the installer level, things are merging as well. The most common type of in-building computer cable for the past several years has been something called Category 5 Cabling. (We will be reviewing this in detail later.) But recently, most telephone installers have been using Category 5 cabling as well, instead of the old-style phone cables. Installation hardware such as punch-down blocks and patch panels are also similar or identical. (Yes, we'll explain what all of these items are as we get further along.)

So, when you see the terms *telecom* or *datacom*, bear in mind that most people use them interchangeably; and if you are unsure about any specific system, find out what type of signal traffic it is designed to carry.

A Brief History

How did we get to where we are now? Well, most of the datacom industry has evolved from telephone technology; and a few of the terms come from the radio industry. None of these technologies are especially difficult, but the terminology is new to most electricians.

Remember that any electronic communication system operates under the same laws that govern power wiring. Ohm's law, Watt's law, Kirchoff's theorems, and circuit calculations are all the same. The applications may be different, but what you learned in apprenticeship school still holds true.

In later chapters we will explain telephone systems in depth; but that is not important to know right now. What is important now, is to know that you will be encountering new terms that come from a different background than the one with which you are familiar. For example, in the color-coding of telephone cables, conductors are identified as *tip* or *ring*. These terms mean nothing to an electrician. In reality, they are simply the positive and negative conductors. Tip is positive, ring is negative. The names come from the old RCA plugs, where one conductor was connected to the tip of the plug, and the other was connected to the ring. There are many more such terms, and each of them will be explained as they are introduced. But do be prepared to learn some new terms.

Signal Transmission

All types of communication systems are concerned with sending signals from one point to another. Whatever form these signals take, our concern is getting them from one place to another, without them being distorted or diminished. To do this economically, special types of conductors, cables, and hardware have been developed.

In general, there are two primary choices for signal transmission methods:
1. Sending electrical signals over copper conductors.
2. Sending pulses of light through optical cables.

There are, of course, a few other options, such as radio signals and microwave signals, but these are used primarily is special circumstances. For now, we will focus on copper and fiber transmission.

There are huge "cable wars" currently being fought between the advocates of copper and fiber. The fiber people say that their product is technically much better, and the copper people say "sure it is, but it costs too much." Which cabling method is better is doesn't matter to us for this discussion: Both types of cabling systems are in wide use; and if you are to be in this industry, you should understand both of them.

Copper Cabling

The copper data cabling used today is unshielded twisted-pair cable (commonly referred to as *UTP*), with four pairs (eight conductors). The conductors are 22 or 24 gauge copper, insulated with some type of

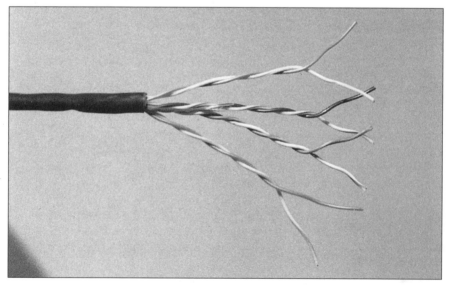

Fig. 1-1. A commonly-used type of UTP cable.

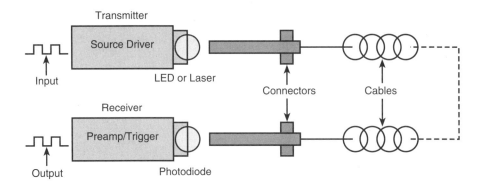

Fig. 1-2. Transmission of optical data.

flouropolymer plastic, and cabled under some type of plastic outer jacket. (See **Fig. 1-1**)

The characteristics of the outer jacket (and sometimes the insulation of the conductors) will vary for one cable rating to another. General duty, riser, and plenum cables will have different types of jackets, due to the fire risks they will encounter in the areas where they are installed.

Copper data cabling is inexpensive, and it is familiar to most of us. But it is limited. First of all, it must be installed very carefully, or it will not transmit signals properly. For example, if these cables are pulled with more than 25 pounds of tension, they tend to untwist, and their transmission characteristics are altered, with the result that the computer network may not work! We'll explain this later, but it is important now to remember that installation is not as easy as you may imagine.

Optical Transmission

Optical fiber systems are used to send all types of data and communication signals. They do this by using pulses of light, rather than pulses of electricity. This is done by sending coded pulses of light into one end of a fiber, and receiving them at the far end, where they are translated and used. (See **Fig. 1-2**)

The light that is used in these systems is an infrared light. It is similar to the light that TV remotes use; and like TV remotes, it is not visible to the

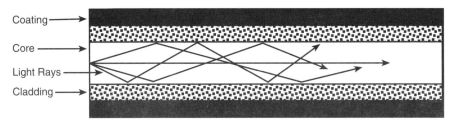

Fig. 1-3. Cross section of an optical fiber.

naked eye. Visible light does not perform nearly as well as infrared in these systems.

The great advantage of optical fiber is that it can handle huge amounts of data; many times as much as copper conductors can. For example, the greatest amount of data sent over copper is usually about 155 Mb/s (*Meg*abits per second). New fiber systems are now sending 40 Gb/s (*Giga*bits per second), and could theoretically go many times that high. As we send more and more data to and from each other, this capacity makes a critical difference.

Networking

A computer network is a collection of electrical signaling circuits, each carrying digital signals between pieces of equipment. There are power sources, conductors, and loads involved in the process. A power source is a network device that transmits an electrical signal. The conductors are the wires that the signal travels over to reach its destination (another network device). The receiver is the load. These items, connected together, make up a complete circuit.

In the computer world, the electric signal transmitted by an energy source is a digital signal known as a pulse. Pulses are simply the presence of voltage and a lack of the presence of voltage, generated in a sequence. These pulses are used to represent a series of zeros and ones (the presence of voltage being a 1, and the absence of voltage being a 0). These zeros or ones are called *bits*. Many years ago, computer engineers began using groupings of eight bits to represent digital "words", and to this day, a series of 8 bits is called a *byte*. These terms are used everywhere in the computer fields.

The key to a successful signal transmission is that when a load receives an electrical signal, the signal must have a voltage level and configuration consistent with what had been originally transmitted by the energy source. If the signal has undergone too much corruption, the load won't be able to interpret it accurately.

A good cable will transfer a signal without too much distortion of the signal, while a bad cable will render a signal useless.

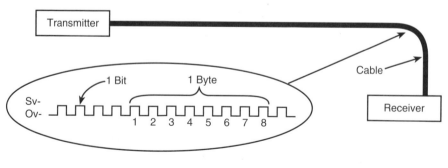

Fig. 1-4. Data signal transmission.

The Limits of Copper Wire

Due to the electrical properties of copper wiring, data signals will undergo some corruption during their travels. Signal corruption within certain limits is acceptable, but if the electrical properties of the cable will cause serious distortion of the signal, that cable must be replaced or repaired.

As a signal propagates down a length of cable, it loses some of its energy. So, a signal that starts out with a certain input voltage will arrive at the load with a reduced voltage level. The amount of signal loss is known as attenuation, which is measured in decibels, or dB. If the voltage drops too much, the signal may no longer be useful.

Attenuation has a direct relationship with frequency and cable length. The higher the frequency used by the network, the greater the attenuation. Also, the longer the cable, the more energy a signal loses by the time it reaches the load.

A signal loses energy during its travel because of electrical properties at work in the cable. For example, every conductor offers some dc resistance to a current (sometimes called copper losses). The longer the cable, the more resistance it offers.

Aside from copper losses (dc resistance, converting some electrical energy into heat), copper cables also have both inductive and capacitive reactance. In an inductive reaction, a current's movement through a cable creates a magnetic field. This field will induce a voltage that will work against any change in the original current. Inductance is measured in Henrys.

Capacitance is a property that is exhibited by two wires when they are placed close together. The electrons on the wires act upon each other, creating an electrostatic charge that exists between the two wires. This charge will oppose change in a circuit's voltage. Capacitance is measured in farads.

Here is an important point to remember - resistance reduces the amount of signal passing through the wires - it does not alter the signal. Reactance, inductive or capacitive, distorts the signal.

Reactance can distort the changes in voltage that signify the ones and zeros in a digital signal. For example, if the signal calls for a one followed by a zero, reactance will resist the switch from voltage to no voltage, possibly causing the load to misidentify what the voltage represents.

Interference

The transmission of a signal can be jeopardized by noise, which can introduce false signals, or noise spikes, at different frequencies. A load may interpret a noise spike as part of a digital signal, distorting the original content of the signal. Common sources of noise spikes include ac lines, telephones, and devices such as radios, microwave ovens, and motors. Some cable testers test for noise, running tests at different frequencies.

Another type of interference is called crosstalk, or more specifically, near-end crosstalk (NEXT). As mentioned earlier, when a current moves through a wire, it creates an electromagnetic field. This field can interfere with signals traveling in an adjacent wire. To reduce the effect of NEXT, wires are twisted-thus the name twisted pair. The twisting allows the wires to cancel each other's noise. The risks of NEXT are highest at the ends of a cable because wire pairs generally don't have twists at their ends, where they enter connectors. If the untwisted end length is too long, NEXT can distort data signals. Also, due to attenuation, signals are strongest when they are transmitted, and weakest when they arrive at their destination. So, the magnetic field of a signal being transmitted from a device through one wire may overwhelm a signal arriving at the same device through an adjacent pair.

NEXT is measured in decibels, which represent a ratio of a signal's strength to the noise generated by crosstalk. The stronger the signal and weaker the noise, the higher the NEXT value. For this reason, *a high NEXT reading is good.* Low NEXT readings, which indicate high crosstalk interference, can mean the cable is terminated improperly.

Two Types of Transmission Problems

Sending data through copper conductors is essentially the same as sending power through copper conductors, except that the amounts of current and the conductors are far smaller; and that the voltage and current characteristics differ.

For power work, we are concerned about the path the current will take, but have fairly little concern for the quality of the power going from one point to another. For data wiring, we must consider two qualities of the transmission:

1. **There must be a clear path from one machine to the next.** Here we are concerned with the signal's strength; it must arrive at the far end of the line with enough strength to be useful.
2. **The signal must be of good quality.** For instance, if we send a square-wave digital signal into one end of a cable, we want a good square-wave coming out of the far end. If this signal is distorted, it is unusable, even if it is still strong.

The main problem we have with power wiring is a loss of power, which we call *voltage drop*. We have the same problem with data signals, when we attempt to send them through conductors with too much resistance (usually due to distance). With data cabling, we call this *attenuation*, and it is virtually the same as voltage drop - not enough power is getting through.

But the problem of signals quality is a completely different concern. We seldom even think about this with power wiring - we simply assume that the current sent through our wiring will be an even sine wave. But with data signals, it is our job (not the power company's) to make sure the signal quality is good. So, getting *enough* signal from one end of the cable to another is one thing we must do; but we must also make sure that the signal at the far end is not too distorted to use.

Pulse Spreading

For data signals to mean anything, they must arrive at the far end of their run resembling the signal that was first put into the cable. The

The Effect Of Long Data Cable Runs On Signal Quality

Fig. 1-5. This is an example of pulse spreading in a long run of data cable. When this happens, the electronic communication circuits can't distinguish between zero and one, making the signal useless.

receivers in data systems require a clear signal; without it, they cannot tell the difference between zeros and ones. Since data signals are always digital (that is, all zeros and ones), a signal that is too distorted means nothing at all.

The most common type of signal distortion is called *pulse spreading*, which is shown in **Fig. 1-5**. Notice that the digital signal sent into the fiber is square. As the signal travels down the fiber, it is distorted, and begins to spread.

Pulse spreading is not a *loss* of signal; it is a *distortion* of the signal. If the pulses spread too much, they will be unintelligible to the receiver, and the communication will be lost.

Categories? Levels? What are they?

The ideas of categories and levels of data cabling began in the late 1980s. With computer networking quickly becoming important, and users needing the ability to send lots of signal (*sufficient bandwidth* in technical terminology), most networks used a *proprietary* type of cable. (*Proprietary* means a type made only by one company, not standardized or industry-wide.)

Manufacturers had found ways of using enhanced telephone wire, called unshielded twisted pair wire (*UTP*), to send data signals. But every manufacturer built and tested their cable differently, and comparing one cable to another was difficult, if not impossible.

Following the widespread acceptance of UTP cable, one distributor, Anixter, began working with manufacturers, *system integrators* (the technical experts responsible for making the entire system work), and large end users to standardize cables for data applications. In 1989, Anixter developed and published a document entitled "Cable Performance Levels", a purchasing specification for communications cables.

Under the original Anixter "Levels" program, there were three levels of cable performance, allowing customers to select the most cost-effective cables for their application. These three levels were:

Level 1: *POTS* - **P**lain **O**ld **T**elephone **S**ervice.

Level 2: Low speed computer terminal and network (ARCNET) applications.

Level 3: Ethernet and 4/16 MB/s Token Ring cabling.

In 1992, two more levels were added, reflecting newer developments in high speed networks:

Level 4: For passive 16 MB/s Token Ring

Level 5: For the copper wire versions of FDDI (Fiber Distributed Digital Interface) at 100 *MB/s* (megabits per second).

In the early 90s, manufacturers turned to the EIA/TIA (Electronic Industries Association/Telecommunications Industries Association) to help create an interoperability standard for building wiring, based on UTP cable. The EIA/TIA committee adopted the Anixter levels 3,4 and 5, calling them *Categories* 3, 4 and 5. These categories have since been adopted internationally. Along with the standards for performance of the components, EIA/TIA participants produced another document, TSB-67, that detailed the testing procedures and specifications that each category must meet.

Of course, in the world of technology, nothing stands still. Currently, new signal transfer methods being introduced. The ones that affect us are *ATM* (Asynchronous Transfer Mode, a new way of sending data in packets, rather than as a continuous stream), *Fast Ethernet*, and *Gigabit Ethernet*. While methods are being used to reduce the bandwidth needs of the network cabling, using encoding schemes like phone modems or multiple pair transmission, it has become obvious that the 100 MHZ limitation on Category 5 cabling is inadequate for the future of UTP.

There are proposals for Category 6 cable, and Category 7 cable. Again, Anixter led the industry into the next stage of defining high performance UTP cable with it's "Levels '97" program. This program updates Cat 5 specifications, and adds Cat 6 and 7 UTP specifications. How quickly Category 6 and Category 7 cables will be formally adopted is a good question, but it will probably happen quickly. Other proposals for higher bandwidth products include shielded or screened twisted pair designs. You can expect these proposed standards to be hotly debated over the next few years. The upgrading of cable alone is not going to provide higher bandwidth performance. It is necessary to develop interconnection hardware that will provide the bandwidth performance wile still being easily field installed.

Chapter 2

Networks

The purpose of networking is to allow a number of computers to operate together, sharing information, and allowing the operator of one computer to read and use programs in another computer.

Here are the fundamental parts of networks:

Connection to the computer: If the network is going to allow other

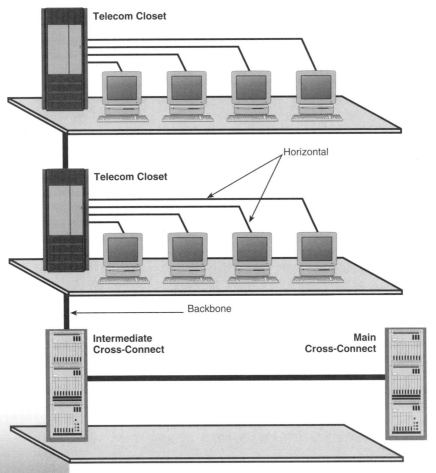

Fig. 2-1. Here, we see the typical data-cabling layout for an office building. The backbone cable shown in this diagram is essentially a "riser" for data. Backbones are the vertical, high-volume cables connecting multiple floors.

computers to access the information in one computer, the network must have some way of getting into it. This is done through the use of a *network interface card.* The network interface card (NIC) is nothing more than a circuit card that fits into one of the expansion slots in the back of a personal computer. It interprets information between the computer and the network, and feeds information in and out of the computer in a way that the computer can accept. This is the most vital link between the processor and the network.

Communication means: Once you have a network interface in place, you need some method of getting the data signals from one computer on the network to another. This can be done in a number of ways. While any of the following methods can be used, the first item on the list (twisted pair cables) is far and away the most common:
 1. Twisted pair cables.
 2. Coaxial cables.
 3. Optical fiber cables.
 4. Radio waves.
 5. Infrared light.
 6. Electronic signals sent through power lines.

The choice of communications means (or *communications media,* as they are sometimes called) is rather important, because of the extremely high frequencies of these signals.

The types of signals that are sent through the network, and the speed at which they are sent, are extremely important details of a network. All parts of the system must be coordinated together to send, carry, and receive the same types of signals. Usually, these details are not something that you have to consider, as long as all parts of your network come from (or at least are specified by) the same vendor. If you ever have trouble with your network, however, you will have to check these signals, and make sure that they are of the correct types. If there are problems with these signals, your network will not function properly. Note that we are not concerned here with the signals getting from one piece of equipment to another, we are concerned with the *type* of signal making its way from one device to another.

Connection pattern: There are several methods for connecting all of the computers on the network together. Some of the most common methods are:
 1. A star pattern.
 2. A ring pattern.
 3. A bus pattern.
 4. A mesh pattern.

These connection patterns are referred to as the network's *architecture,* or the *topology.*

In addition to these connection patterns, there are also others which are less-commonly used. Among these are the tree structure (a group of stars, connected to a bus), and a star-ring.

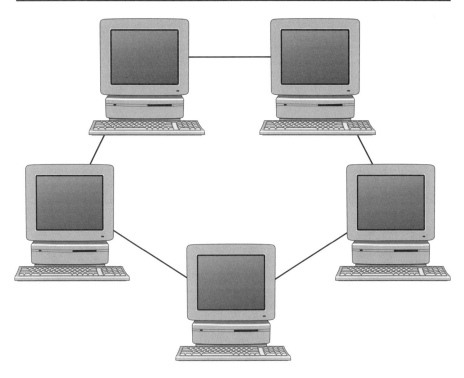

Fig. 2-2. A ring pattern topology.

Which connection pattern you need to use is determined by the brand of network you plan to install.

Ring Topology

A ring topology consists simply in connecting the computers together in a large ring. This is done with either fiber optic, twisted pair or coaxial cables.

Note that in this arrangement, each computer has two cables connecting to it, one from the previous computer, and one to the next computer in line. When you draw this arrangement, it looks like a very simple ring; but in actual practice, it is more like a chain run from computer to computer, with a final run back to the first computer, once the end of the line has been reached.

Star Topology

A star topology is an arrangement where one computer is located at the center of the "star", this is the file server, or main computer. From this central computer, there are a number (exactly how many depends on the type of network, and the application) of cable runs to the various satellite stations (nodes). In actual practice each computer has only one cable

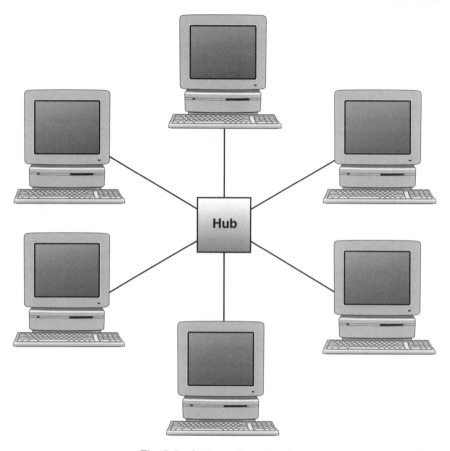

Fig. 2-3. A star pattern topology.

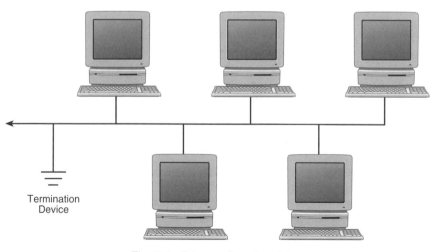

Fig. 2-4. A bus pattern topology.

connected to it, with all of these cables feeding into one central location, where the file server is located.

Bus Topology

On a diagram, a bus topology looks like all of the computers are connected to one central cable; but in actual practice, it is somewhat different. In an actual installation, a bus topology requires two cables run to most of the computers (they "splice through" internally, making them continuous). The computers at the end of the line, however, only have one cable connecting to them. If you are not aware of this fact, it can be very confusing when you are trying to figure out what type of network is installed in an office where you are working.

Attenuation

Attenuation is the loss of signal power. An attenuating signal is a signal that is weakening. This attenuation is normally measured as a number of decibels per 100 feet, at a given frequency. For LAN (Local Area Network) work, attenuation is measured in decibels over the link, or decibels per 100 meters.

Signals sent over copper wires deteriorate differently at different frequencies - the higher the frequency, the greater the attenuation. These losses come primarily from the capacitance of the cable, the inductance of the cable, or from *copper losses* (simple resistance, resulting in heat). Attenuation is a problem, since a weakened signal can only be picked up by a very sensitive receiver. Such very sensitive receivers are quite expensive; therefore, low attenuation is desirable.

Signal attenuation also depends on the construction of the cables, particularly the dielectric characteristics of the cable insulation. For example, you could have two 100 ohm, 24-gauge cables; and one of the cables might have a lower impedance than the other, strictly because of the construction characteristics of the cables. This would allow the cable with lower impedance to be used over longer distances with better results.

Cable Impedance

Impedance, which is the total opposition to current flow, is an important consideration for coaxial and twisted pair cables for computer systems. Cables used in computer networks are rated by their impedance characteristics. For example, the cable used in the example in the last section was specified as a 100-ohm cable. That means that the cable has a *characteristic impedance* of 100 Ohms.

Characteristic impedance is another confusing telecom term. Characteristic impedance refers to the internal signal-transmission characteristics of the cable. You cannot measure the characteristic impedance of a cable with an ohmmeter, and the number does not refer to a number of ohms per foot or per hundred feet. The really important thing with the characteristic impedance of cables is that you never switch cable with differing impedances. If you are running 100-ohm cable, you cannot insert a piece of 150-ohm cable in the network. If you were to do this, the signal would tend to reflect off of the 100-to-150-ohm junction, ruining the transmission. In short, the link will not pass signal if you mix cables of different impedance.

A more scientific explanation is as follows: Characteristic impedance is important because it determines the amount of power that can be transferred between devices. Whenever two electrical devices are connected, the power transferred from one to the other is maximized when their impedances are equal. (More correctly, when the output impedance of the driving device is equal to the input impedance of the driven device.) Any amount of impedance mismatch will cause some of the power to be reflected back into the source device. This is why cables of differing impedance should never be directly-connected, and why impedance-matching transformers are used. An unterminated piece of coax is effectively attached to an infinite impedance (being an open circuit), which is the ultimate mismatch; and this causes all the power arriving at the open end to be reflected. The reflections look like collisions to transmitting devices and usually bring the network to a quick halt.

There are a few cases where very short runs of cable can be mismatched, and the network not crash. A device connected to a transmission line sees only the electrical characteristics of the cable, not the device on the other end. The rule of thumb is that a conductor is a transmission line when its propagation time (the time it takes a signal to travel from one end to the other) is at least as great as the rise time of the signal being applied. You can sometimes get away with using the wrong kind of cable, but only in very short pieces.

Capacitance

As we mentioned, *capacitance* is the most common element of attenuation. This capacitance usually comes in the form of *mutual capacitance,* which is capacitance between the conductors within the cable. Here is why capacitance is a big factor: Data transmissions are square wave signals. When the capacitance is too great, it has a tendency to round-off the data signals. This results in pulse spreading.

When data signals are have a clear square wave, they are easily intelligible to the receiver. If, however, the signals become rounded, they can be confusing to the receiver, causing malfunctions.

Cross-talk

Cross talk is the amount of signal that is picked up by a quiet conductor (a conductor with no signal being transmitted over it at the moment) from other conductors that are conducting data. This signal is picked up through electromagnetic induction, the same principal by which transformers operate. Cross talk contaminates adjacent lines, and can cause interference, overloaded circuits, and other similar problems.

Cable Shielding

The most common method to prevent cross talk and electromagnetic

interference is to place a shield around the conductors. The three principal types of shields are these:
1. Longitudinally-applied metallic tape.
2. Braided conductors, such as are commonly used in coaxial cables.
3. Foil laminated to plastic sheets.

Spirally-applied metallic tapes can be used also, but they are not generally good for data transmission. Shielding stops interference and cross talk by absorbing magnetic fields. Since the shielding is conductive itself, when the magnetic field crosses through it, it is absorbed into the shield. It does induce a current into the shield, but since this current is spread over the wide, flat surface of the shield, it is diffused, and is not usually strong enough at any one point to cause a problem.

Foil shields will usually reduce electromagnetic interference by 35 decibels, wire braid shields will generally reduce this type of interference by 55 decibels, and a combination of the two types of shields will reduce interference by over 100 decibels.

It is also possible to use filters to reduce interference, although they must be changed every time the network's data transmission rate is altered.

Shielded cables are not the most typically-used type for structured cabling systems - unshielded twisted-pair (UTP) cables are usually used, and shielded cables are used mainly when there may be a problem with electromagnetic interference.

Common Types of Network Cabling

Unshielded twisted pair cables, 22-24 gauge (UTP)
    Advantages - Inexpensive; may be in place in some places; familiar and simple to install.
    Disadvantages - Subject to interference, both internal and external; limited bandwidth, which translates into slower transmissions. Somewhat vulnerable to security breaches; may become obsolete quickly because of new technologies.

Shielded twisted pair cables, 22-24 gauge (STP)
    Advantages - Easy installation; reasonable cost; resistant to interference; better electrical characteristics than unshielded cables; better data security; easily terminated with modular connector.
    Disadvantages - May become obsolete due to technical advances; can be tapped, breaching security.

Coaxial cables
    Advantages - Familiar and fairly easy to install; better electrical characteristics (lower attenuation and greater bandwidth) than shielded or unshielded cables; highly resistant to interference; generally good data security; easy to connect.
    Disadvantages - May become obsolete due to technological advances; can be tapped, breaching security.

Optical fiber cables
    Advantages - Top performance; excellent bandwidth (high in the gigabit

range, and theoretically higher); very long life span; excellent security; allows for very high rates of data transmission; causes no interference and is not subject to electromagnetic interference; smaller and lighter than other cable types.
Disadvantages - Slightly higher installed cost than twisted-pair cables.

Wireless Transmission

In the past several years, there has been a good deal of interest in wireless networks. The reasons for this interest is due to the expense of installing and moving cables (the installation expenses are usually far higher than the material expenses), and the frequent moving of terminals within a large office. With a wireless network, no data transmission cables are required. Additionally, within the range of the radio signals, a terminal can be moved anywhere.

These wireless networks are usually somewhat more expensive than cabled networks, at least from a material cost standpoint. They are often more expensive from an initial cost standpoint as well. Where they really save money is when terminals must be moved from one spot to another. Money can also be saved in locations where it would be especially difficult to install cables.

The number one problem with wireless networks is their speed. It is very difficult to transmit data over available radio frequencies as fast as can be done over copper wires. To compensate for this impediment, some wireless networks have gone to multiple frequencies. By sending signals over several radio frequencies at the same time, signals can be sent far faster than over a single frequency. Special software is required to send these signals in a coded format.

Infrared Transmission

Another method of transmitting data signals without the use of wires is by using infrared light. By sending pulses of infrared light in the same patterns as electronic pulses sent over cables, it is possible to send data from one place to another.

Infrared light is used for this purpose (rather than other types of light), because it is invisible to the naked eye, and because it is inexpensive to implement. This is a variation on the same technology we use for TV remote controls. The distance between terminals is normally limited to around 80 feet, although newer systems exceed that figure.

This technology works fairly well, although there are problems that develop in offices with numerous walls. Just like normal light, infrared light cannot pass through walls. In open offices, this is not much of a problem, but in walled-off offices, remote transmitter/receivers are often required. To do this, a transmitter/receiver is mounted in an area where it will easily receive data signals, and it is connected to the computer with a cable. Not all of these systems use combination transmitter/receivers, however; some of the use separate transmitters and receivers.

As with other wireless signal transmission, infrared networks require all of the parts that conventional networks require. Where the data cables would normally connect to the back of the computer, however, transmitter/receivers are installed instead. Depending on the design, the units can be rather large and awkward.

These infrared signals can be used to send and receive through these devices, usually at a rate of between 4 and 16 million bits per second.

One great advantage of the wireless networks is that they can be set up anywhere, almost instantly. This can be a terrific feature for people whose work location is constantly shifting from one place to another. Some people in the trade have come to call these systems "network in a box".

Powerline Carrier Networks

A new method of sending data signals involves sending them through regular power lines. The method by which this is accomplished is as follows:
1. A special device is connected to the computer in place or network cable.
2. Data from the computer goes to this device, and are modified.
3. The device (which is simply plugged into a wall outlet) sends these modified signals into the building's wiring system.
4. The signals are received by other electronic devices that are connected to the other computers on the network.
5. All of the electronic devices "read" the data signal, but only the computer to which it is sent will have the data sent to it from its receiver (the electronic device).

This system is very attractive from the standpoint of installation costs. No cables are required, and the installation is very quick and easy. The system superimposes data signals right over the power line voltage. The voltage and frequency differences are very easily separated from each other by tuned receivers. (Power lines operate at 60 cycles per second, and 120 volts. Data signals are normally in the 5 volt range, and have a frequency of hundreds of thousands or millions of cycles per second.)

Powerline carrier systems will never meet the technical performance of optical fiber cables, or even other types of cables. They are, however, a very appealing option for low-speed systems.

The installation of these systems is not completely hassle-free, however. Powerlines do not travel unbroken throughout whole buildings, or even through parts of buildings. To bridge the gaps in wiring systems, special devices called "signal bridges" are required. These devices connect to two separate wiring systems, and transfer the data signals from one to another, without allowing current to be transferred from one system to another (which would cause major problems, and likely injuries). Signal amplifiers are also frequently required when wiring systems cover long distances. There can also be filters and other devices required for these systems, depending on the location of the installation.

Other Transmission Means

While the methods of transmission that we have gone through are by far the most common methods, there are other methods that are sometimes used. The chief among these are microwave and laser signal transmission.

Both of these methods have limited circumstances in which they are applicable, but can be very effectively used where they are appropriate.

The best and most common application of these technologies is sending signals between two buildings that have a clear line of sight between them. In other words, that there are no objects blocking a path between the two buildings. The typical method is to set up transmitter/receivers on the roofs of both buildings, and to connect both ends to the networks in their buildings. The buildings can be up to about one kilometer apart for most systems, and even further for others. Signal transmission rates of 1.5 Mb/s (megabits per second) or greater are not uncommon.

It is important to remember in planning such a system that the management of any buildings that are under the transmission area should be notified, and any appropriate permissions granted before any work is begun.

Inexpensive Cables for High-speed Networks

Normally, wire-pair cables cannot carry high bandwidth signals fast enough for network communications. The wires also radiate like antennas, so they interfere with other electronic devices. However several developments allowed the use of simple wire for high speed signals.

The first development was the use of twisted pair cables. The two conductors are tightly twisted (1-3 twists per inch) to couple the signals into the pair of wires. Each pair of wires is twisted at a different rate to minimize cross-coupling.

The next step was to use *balanced transmission* to minimize electromagnetic emissions. Balanced transmission works by sending equal but opposite signals down each wire. The receiving end looks at each wire and sees a signal of twice the amplitude carried by the pair of wires; this helps for getting enough signal through the link. While each wire radiates because of the signal being transmitted, each wire carries the opposite signal so that the two wires cancel out the radiated signals and reduce electromagnetic interference.

Using these techniques, and having two pairs of wires sending signals in opposite directions, it has been possible to adapt unshielded twisted pair (UTP) cables to work with Ethernet and token Ring networks. At higher speeds, signals are compressed and encoded and multiple pairs are used for transmission in each direction to allow operation with networks of over 100 megabits per second.

Typical UTP cables have four twisted pairs in the cable, two of which are used to simultaneously transmit signals in opposite directions, creating a full duplex link. Some of the higher speed networks now use all four pairs to reduce the total bandwidth requirement of any single pair.

More complex cables have also been used for LANs, including shielded twisted pair (STP) cable with each pair of wires shielded individually and an overall shield provided, and screened twisted pair cable (ScTP) that has four twisted pairs inside a foil shield.

Ethernet

Ethernet is generally the least expensive high speed LAN system, and is by far the most commonly-used. Ethernet adapter cards for a network computer usually range in price from $60 to $120. They transmit and receive data at speeds of 10 million bits per second through up to 100 meters (90 meters horizontal, and 10 meters in closet and wall) of cable to a *hub* device normally stacked in a wiring closet. The hub adds about $50 to the cost of each desktop connection. You will notice that Ethernet and 568 standards are the same - Ethernet is designed to run over the 568 structure.

Standards

The early development of Ethernet was done by Xerox research. The name *Ethernet* was a registered trademark of Xerox Corporation. The technology was refined and a second generation called Ethernet II was widely used. Ethernet from this period is often called DIX after its corporate sponsors Digital, Intel, and Xerox. As the holder of the trademark, Xerox established and published the standards.

Obviously, no technology could become an international standard for all sorts of equipment if the rules were controlled by a single US corporation. So, the IEEE was assigned the task of developing formal international standards for all Local Area Network technology. It formed the "802" committee to look at Ethernet, Token Ring, Fiber Optic, and other LAN technology. The objective of the project was not just to standardize each LAN individually, but also to establish rules that would be global to all types of LANs so that data could easily move from Ethernet to *Token Ring* (another popular type of network).

The IEEE eventually published a set of standards. The most important of these are:
- 802.3 - Hardware standards for Ethernet cards and cables
- 802.5 - Hardware standards for Token Ring cards and cables
- 802.2 - The new message format for data on any LAN

Collisions

Ethernet uses a protocol called CSMACD. This stands for *Carrier Sense, Multiple Access, Collision Detect*. The *Multiple Access* part means that every station is connected to a single set of wires that are connected together to form a single data path. *Carrier Sense* means that before transmitting data, a station checks the conductors to see if any other station is already sending something. If the LAN appears to be idle, then the station can begin to send data.

A standard Ethernet station sends data at a rate of 10 megabits per second. (Some newer systems transmit at higher speeds.) After the electric signal for the first bit has traveled about 100 feet down the wire, the station has begun to send the second bit. However, an Ethernet cable can run for hundreds of feet. If two stations are located 250 feet apart on the same cable, and both begin transmitting at the same time, then they will be in the middle of the third bit before the signal from each reaches the other station.

This explains the need for the *Collision Detect* part. Two stations can begin to send data at the same time, and their signals will collide nanoseconds later. When such a collision occurs, the two stations stop transmitting, and try again later after a randomly chosen delay period.

Devices

Here are a few of the devices that are commonly used in Ethernets:
- A **repeater** receives and then immediately retransmits each bit. It has no memory and does not depend on any particular protocol. It duplicates everything, including the collisions.
- A **bridge** receives the entire message into memory. If the message was damaged by a collision or noise, it is discarded. If the bridge knows that the message was being sent between two stations on the same cable, it discards the message. Otherwise, the message is queued up and will be retransmitted on another Ethernet cable. The bridge has no address. Its actions are transparent to the client and server workstations.
- A **router** acts as an agent to receive and forward messages. The router has an address that is known to the client or server machines. Typically, machines send messages directly to each other when they are on the same cable, and they send messages addressed to another zone, department, or sub-network to the router. Routing is a function specific to each protocol. For example, on an IPX system, the Novell server can act as a router. For SNA, an APPN Network Node does the routing. TCP/IP can be routed by dedicated devices, UNIX workstations, or OS/2 servers.

Problems

Ethernets fail in three common ways:
1. A nail or other object can break one of the conductors, causing an open circuit.
2. A screw or other object can touch one or more of the conductors and short them to an external grounded metal shield, conduit, or other grounded metal, attenuating the signal.
3. A station on the network can malfunction and start to generate a continuous stream of electronic noise, thus blocking legitimate transmissions.

A time domain reflectometer is used to find problems in an Ethernet. It plugs into any attachment point in the cable, and sends out its own voltage pulse. The effect is similar to a sonar ping. If the cable is broken, then there is no proper terminating resistor. The pulse will hit the loose end of the broken cable and will bounce back. The test device senses the echo, computes how long the round trip took, and then reports how far away the break is in the cable.

If an Ethernet cable is shorted out, a simple voltmeter determines that the proper resistance is missing from the signal and shield wires. Again, by sending out a pulse and timing the return, the test device can determine the distance to the problem.

Newer generations of "smart" hubs can perform part of the error detection and reporting function. For example, they could isolate a problem in the connection to a particular desktop workstation and automatically isolate that unit from the rest of the network.

Frames

A block of data transmitted on the Ethernet is called a *frame*. The first 12 bytes of every frame contain the 6 byte destination address (the recipient) and a 6 byte source address (the sender). Each Ethernet adapter card comes with a unique factory installed address (called a *universally administered address*). Use of this hardware address guarantees a unique identity to each card.

The source address field of each frame must contain the unique address (universal or local) assigned to the sending card. The destination field can contain a *multicast* address representing a group of workstations with some common characteristic.

In normal operation, an Ethernet adapter will receive only frames with a destination address that matches its unique address, or destination addresses that represent a multicast message. However, most Ethernet adapters can be set into "promiscuous" mode where they receive all frames that appear on the network. If this poses a security problem, a new generation of smart hub devices can filter out all frames with private destination addresses belonging to another station.

There are three common conventions for the format of the remainder of the frame:
1. Ethernet II or DIX
2. IEEE 802.3 and 802.2
3. SNAP

We will not go through the technical details of each protocols, since they are not important as they apply to installations.

How Ethernets Are Used

Ethernet was originally supposed to be a single common medium with multiple connections. That may be true for older installations and laboratory environments. However, new desktop installations bring Ethernet to the desktop over category 5 cables and frequently build the backbone with optical fiber cables.

The connection between the hub in the wiring closet and the adapter card in the PC forms a single point-to-point Ethernet segment between two stations. The connection to the rest of the network involves electronics in the hub. In current use, this is done with a repeater that copies every bit and allows collisions.

A new generation of even smarter hubs provides a "bridge" connection between the main backbone and the desktop. Only multicast messages and private messages specifically addressed to the PC are forwarded to the desktop. This has two advantages:
1. It provides greater security, because the desktop user cannot spy on traffic addressed to other nodes.
2. It provides each desktop user with an isolated, private 10 megabit data path free of collisions.

The connection between hubs can then use a higher speed fiber optic protocol (such as ATM) to deliver much greater performance than simple Ethernet. This hybrid (Ethernet to the desktop, fiber between the hubs) represents a compromise of high performance and low cost.

However, bridging Ethernet to any other LAN protocol requires some attention to frame formats. Bridges must be aware of the protocol conventions and select the correct frame format when moving data onto or off of an Ethernet.

Other Network Terms

Baseband network. A baseband network is one that provides a single channel for communications across the physical medium (cable), so that only one device can transmit at a time. Devices on a baseband network, such as Ethernet, are permitted to use all the available bandwidth for transmission, and the signals they transmit do not need to multiplexed onto a carrier frequency. An analogy is a single phone line: Only one person can talk at a time—if more than one person wants to talk everyone has to take turns.

Broadband network. A baseband network is in many ways the opposite of a baseband network. With broadband, the physical cabling is virtually divided into several different channels, each with its own unique carrier frequency, using a technique called *frequency division modulation*. These different frequencies are multiplexed onto the network cabling in such a way as to allow multiple simultaneous "conversations" to take place. The effect is similar to having several virtual networks going through a single piece of wire. Network devices tuned to one frequency can't hear the signal on other frequencies, and visa-versa. Cable-TV is the best example of a broadband network, with multiple conversations (channels) transmitted simultaneously over a single cable; you pick which one you want to see by selecting the frequencies.

OSI Model. The *Open Systems Interconnect* (OSI) reference model is the ISO *(International Standards Organization)* structure for network architecture. This Model outlines seven areas, or layers, for the network. These layers are (from highest to lowest):
 7. *Applications:* Where the user applications software lies. Such issues as file access and transfer, virtual terminal emulation, interprocess communication and the like are handled here.
 6. *Presentation:* Differences in data representation are dealt with at this

level. For example, UNIX-style line endings (CR only) might be converted to MS-DOS style (CRLF), or EBCIDIC to ASCII character sets.
5. *Session:* Communications between applications across a network is controlled at the session layer. Testing for out-of-sequence packets and handling two-way communication are handled here.
4. *Transport:* Makes sure the lower three layers are doing their job correctly, and provides a transparent, logical data stream between the end user and the network service. This is the lower layer that provides local user services.
3. *Network:* This layer makes certain that a packet sent from one device to another actually gets there in a reasonable period of time. Routing and flow control are performed here. This is the lowest layer of the OSI model that can remain ignorant of the physical network.
2. *Data Link:* This layer deals with getting data packets on and off the wire, error detection and correction, and retransmission. This layer is generally broken into two sub-layers: The LLC (Logical Link Control) on the upper half, which does the error checking, and the MAC (Medium Access Control) on the lower half, which deals with getting the data on and off the wire.
1. *Physical:* The nuts and bolts layer. Here is where the cable, connector and signaling specifications are defined.

10Base5, 10BaseT, 10Base2, 10Broad36, etc. These are the IEEE names for the different physical types of Ethernet. The "10" stands for signaling speed: 10MHz. "Base"means Baseband, "broad" means broadband. Initially, the last section was intended to indicate the maximum length of an unrepeated cable segment in hundreds of meters. This convention was modified with the introduction of 10BaseT, where the T means twisted pair, and 10BaseF where the F means fiber.

In actual practice:

10Base2   Is 10MHz Ethernet running over thin, 50 Ohm baseband coaxial cable. 10Base2 is also commonly referred to as thin-Ethernet.

10Base5   Is 10MHz Ethernet running over standard (thick) 50 Ohm baseband coaxial cabling.

10BaseF   Is 10MHz Ethernet running over fiber-optic cabling.

10BaseT   Is 10MHz Ethernet running over unshielded, twisted-pair cabling.

10Broad36 Is 10MHz Ethernet running through a broadband cable.

Driver. The software that allows an Ethernet card in a computer to decode packets and send them to the operating system and encode data from the operating system for transmission by the Ethernet card through the network. By handling the nitty-gritty hardware interface chores, it provides a device-independent interface to the upper layer protocols, thereby making them more universal and easier to develop and use.

Structured Cabling & Ethernet

The first computer networks were proprietary - that is, every manufacturer's system was different. You ran cable one way for an IBM system, an entirely different way for a DEC system, and a third way for an AT&T system. And not only did cable routes vary, but cable types varied. There were lots of manufacturers doing this, and installers were generally confused. Eventually, the manufacturers got together (through the EIA/TIA organization), and came up with a generic cable routing pattern that they

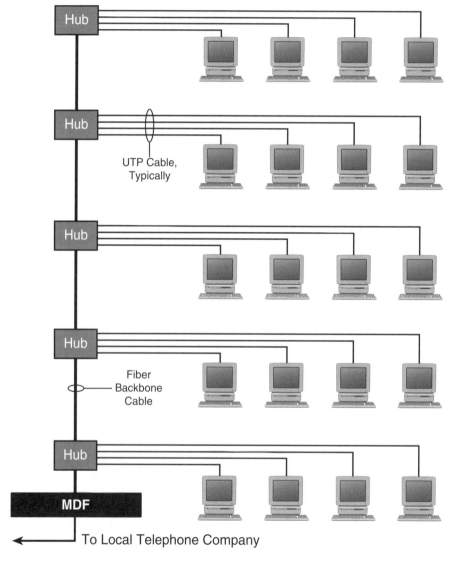

Fig. 2-5. Typical structured cabling layout (568).

would all use - a standard for cabling buildings. They called this new standard EIA/TIA 586, and gave it the name *Structured Cabling*. 586 A is now the most commonly-used version of the standard, and the one that almost all computer networks in buildings now follow.

Structured cabling generally refers to a network cabling system that is designed and installed according to pre-set standards. The benefits of structured cabling are:

1. Buildings, new or re-furbished, are pre-wired without needing to know future occupant's data communication needs.
2. Future growth and re-configuration are accommodated by pre-defined topologies and physical specifications, such as distances.
3. Support of multi-vendor products, including cables, connectors, jacks, plugs, adapters, baluns, and patch-panels, is assured.
4. Voice, video and all other data transmissions are integrated.
5. Cable plant is easily managed and faults readily isolated.
6. All data cabling work can be accomplished while other building work is underway.

Following are the standard materials used for a structured cabling (568) system:

Four-pair 100 ohm UTP cables. The cable consists of N⁰ 24 AWG thermoplastic insulated conductors formed into four individually twisted pairs and enclosed by a thermoplastic jacket. Four-pair, N⁰ 22 AWG cables that meet the transmission requirements or four-pair, *shielded* twisted pair cables that meet the transmission requirements may also be used. To assure a minimum of crosstalk, the pair twist rate of any single pair must not be exactly the same as that of any other pair.

Network Architectures

There are two basic types of computer networking models: *Centralized* computing and *Client/Server* computing.

Centralized Computing: In the past, corporate data communications involved accessing a central computer. Everybody went to this one

Network	IEEE802.3 FOIRL	IEEE802.3 10baseF	*IEEE802.5 Token Ring	ANSI X3T9.5 FDDI	ESCON IBM
Bitrate (MB/s)	10	10	4/16	100	200
Architecture	Links	Star	Ring	Ring	Branch
Fiber Type	MM, 62.5	MM, 62.5	MM, 62.5	MM/SM	MM/SM
Link Length (km)	2	—	—	2/60	3/20
Wavelength (nm)	850	850	850	1300	1300
Margin (dB, MM/SM)	8	—	12	11/27	8*(11)/16
Fiber BW(MHz-km)	150	150	150	500	500
Connector	SMA	ST	FDDI	FDDI	ESCON

*IBM specifies a nonstandard method of testing cable plant loss that reduces the loss to 8 dB max. However, the component specifications are simmilar to FDDI< so testing to FDDI margins is appropriate.

Table 2-1. Fiber optic datacommunications networks.

computer to take care of a particular task or business process. Input to the computer was made using interactive (dumb) terminals. Later, smart terminals provided for batched input to the mainframe. These types of terminals are often found in retail chains where stores download sales information to the mainframe at the end of the day.

Client/Server Computing: The general availability of microprocessor-based Personal Computers changed the way networking was done. With more intelligent terminals, most, or in some cases all, of the processing load is performed at the desk through a Personal Computer, and not at the mainframe.

Along with the Client/Server computer model came new methods of getting computers to talk to one another. A high-speed transmission media was needed, called a Local Area Network (LAN). Also, computers had to talk the same network language to form a Network Operating System (NOS). With a NOS, a computer's operating system is integrated into the network. These systems have been developed and made widely available by companies such as DEC, IBM, Novell, LanTastic, 3Com, Xerox, Banyan-Vines, and Microsoft.

The Centralized computing model has the following attributes:
- Relies heavily on WAN (wide area network) technologies
- Well-suited for mission-critical information
- High computer cost
- Low end-user equipment costs
- Lower network management costs
- Higher transmission facility costs
- Lacks flexibility and customization

The Client/Server computing model has the following attributes:
- Relies on both LAN and WAN technologies
- Flexible deployment - easily customized
- Low computer cost
- Increased end-user equipment costs
- Lower transmission facility costs
- Increased network management costs

Both of these *architectures* (network structures) are likely to be found within a typical company's network, especially a large company. Companies with both types of architectures must have hardware and software that can adapt quickly and easily to either model, without excessive loss of data and productivity. At the time of this writing, mainframe and facilities costs are decreasing, resulting in more justifiable use of centralized computing.

Common Architectures

While there are probably hundreds of network topologies that have been used over recent years, we will boil them down to three:

Small, single segment networks. Usually 100 or fewer users. The common topologies are Ethernet and Token-Ring. The network usually has no more than two or three servers.

Medium-sized, collapsed backbone networks. Usually 1,000 or fewer users. Common topologies are Ethernet and Token-Ring. A single router, or just a couple routers, sit as the backbone of the network and provide connectivity for the network. The network usually has ten or so servers.

Large, high-speed connected networks. Usually more than 1,000 users and often involving more than one building or a number of floors of a large office building. Desktop connectivity is still with Ethernet and Token-Ring, backbone is most often FDDI. Routers sit on the high-speed backbone and provide connectivity to attached Ethernet and Token-Ring segments.

Physical Structure

You should plan your cable plant around two levels of distribution. This will accommodate changes in office layout, occupants and locations of equipment. The *main distribution frame* (MDF) will probably be co-located with primary network services, bridging and routing, and possibly network management. The MDF will be connected by riser cabling to wiring closets, or *intermediate distribution frame* (IDFs), which support workstation locations in clearly defined geographic portions of the building. Hopefully, the IDFs will be stacked above each other on the various floors, to make it easy to run riser cabling connecting the IDFs to your MDF.

You will want to wire each workstation location supported by an IDF directly back to a patch panel that supports a hub (data) and to a punch

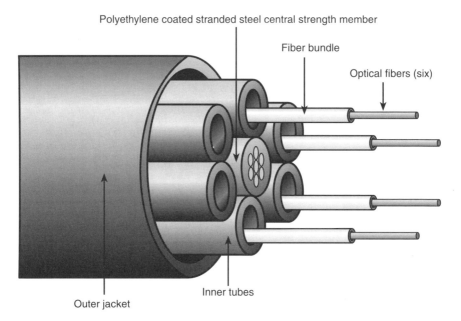

Fig. 2-6. Six tube LAP cable construction. The tube cable, used in an ABF, comes with factory-installed inner tubes, each accommodating a single-fiber bundle. A fiber bundle contains 2 to 18 optical fibers—either single-mode or multimode construction.

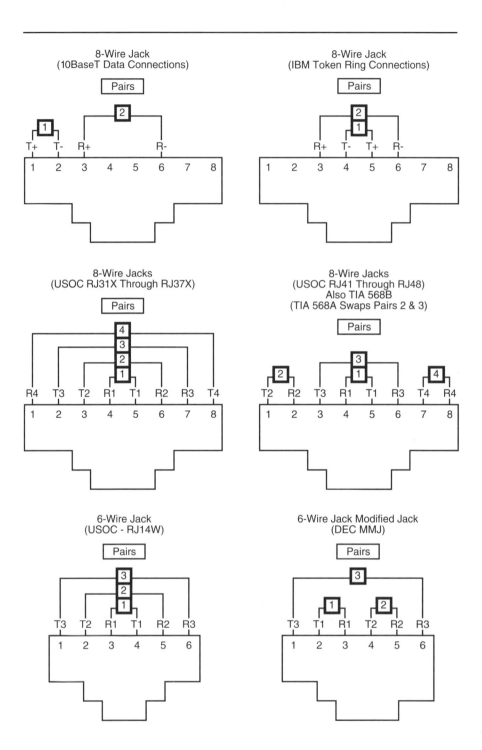

Fig. 2-7. Common unshielded twisted pair data and voice wiring schemes.

block (voice) in that IDF. This "star" wiring configuration lends itself to any network topology. And by designing the voice cabling to run parallel with the data cabling, and including spare cable runs to each workstation, it's possible to run any service on any wire from any IDF built into your design.

When planning your cable plant layout, resist the temptation to exceed the 90-meter length specified by the EIA/TIA standard (total electrical length of the station cable, from IDF patch panel to wall connector) for your station cabling, even if your choice of hubs will drive the signal farther than that on Cat 5 UTP. Your next generation of hubbing equipment, designed to that standard, may not.

There is a wide range of media choices available, but a modern network is typically driven by bandwidth and constrained by cost. This narrows to a choice, for most networks, between multimode fiber and Cat 5 UTP. While nearly everyone agrees that fiber to the desktop is the superior system, most settle for fiber only in the backbone. The extra cost of FO hubs and network interface cards (NICs) has historically been unjustifiable (though prices have been dropping recently).

Most networks run at either 10 Mbps or 16 Mbps. If you have 30 nodes on a segment, that 10 or 16 Mbps is shared among perhaps 30 nodes. Each node receives 1/30 of 10 Mbps or, in reality, 1/30th of 25 percent of 10 Mbps-about the maximum traffic for an Ethernet segment. That is about 83 Kbps per node. But trends are moving rapidly toward switched rather than shared media. Replace the hub with a switch and each node gets a full 10 Mbps. Replace the 10-Mbps cards with 10/100-Mbps cards and each node gets a full 100 Mbps.

If the switch supporting an IDF has 90 workstations attached, the riser between that IDF and your MDF will have to handle the network traffic for all 90. If your planning factor for station traffic is 100 Mbps to each workstation, then your riser needs 100 Mbps or better. If your future may require some standard above 100 Mbps, say 155 Mbps or 622 Mbps, then fiber may be the right choice for your riser. However, if you have installed spare floor sleeves for future new riser cabling, you can always add the additional riser capacity when you need. Note that station cabling is another matter altogether. The last thing you want to do is pull more station cabling midway through the life of your cable plant.

One interesting option for the riser is *Air Blown Fiber* (ABF). Although ABF is a proprietary system, the fiber strands will support a standards-based installation. The fiber bundles are blown through pre-installed hard rubber tubes with compressed air or Carbon Dioxide. You won't need to pull a lot of excess fiber in the initial installation because it is quite simple to blow additional fiber through the pre-installed tubes as your bandwidth requirements increase.

Now that we have a very cost-effective solution for data, what about voice service? You will need a cable plant for voice that parallels your data cable plant. The voice MDF will be located near the data MDF and feed the same riser path. While the data riser will contain relatively few cables to each IDF, the opposite is true for voice service. The voice riser will need either two pair or four pair for each telephone instrument, all the way from the MDF to the desktop. Since each wire pair will need to support only 56 kbps to 128 kbps, there is little need for using any media higher than Cat 3 UTP. (Nonetheless, a lot of people use Category 5 cabling anyway.)

Voice station cabling, however, is another matter. One highly effective design strategy is to run three or four Cat 5 cables from the IDF closet to each workstation and terminate them all in the same information outlet box with RJ-45 jacks. That would allow you to use any of the four cables for voice, data, printer, modem or fax. The difference between cables at a workstation outlet depends entirely upon whether they terminate at the IDF on the data patch panel, or the voice punch blocks.

Other Media Choices. Multimode fiber and Cat 5 typically is the best combination of bandwidth and cost. Yet there may be other reasons for choosing fiber as a medium. Because the photons conducted by fiber neither emit an electromagnetic field nor are effected by such a field, data carried by fiber is free from Radio Frequency Interference (RFI) and Electromagnetic Interference (EMI), and it is secure from most attempts to steal your data. However, the EMI/RFI over copper issue is easily dealt with by observing minimum distances from sources of RFI/EMI (specified in the EIA/TIA standard) and by running UTP in conduit for shielding past particularly "dirty" emitters.

Another property of fiber is its ability to carry data long distances. Standard 62.5/125 micron multimode fiber (FDDI grade) is generally used for distances of up to 2 kilometers. Single mode fiber, used by long distance carriers, can transmit data up to 80 kilometers without signal regeneration, and can handle data rates of many gigabits per second (OC-48 is 2.488 Gbps). It's unlikely that electro-optics for data rates in the gigabit range will become cost effective for LAN implementation in the next few years.

Coaxial cable offers longer transmission distances and better freedom from EMI/RFI than does UTP, and the implementation is less expensive than fiber, but coaxial does not lend itself to the "star-wired" physical topology that is basic to a structured cable plant. Coaxial's only valid use is to link a stack of remote workgroup hubs with an IDF at some distance, when cost of implementation is the driving issue. Still, fiber is a far better choice, and the price of the implementing electronics is rapidly becoming competitive with those required for coaxial.

Cat 3 or Cat 4 UTP is another option. Cat 3 will serve well for today's 10-Mbps LANs, and Cat 4 is fine for a 16-Mbps installation, but neither is suited for higher-speed networks. Given the slight price difference between Cat 3 and Cat 5, you should never plan to install less than Cat 5 for any new wiring plant unless you can be certain that it will only be used for voice during the life of the installation.

Other Design Issues

You must plan for a grounding backbone that runs parallel to your communications riser and is tied into building steel and a buried grounding electrode. This is so important that the EIA/TIA has established a separate standard (Standard 607) for the Telecommunications Bonding Backbone.

The path from IDF to workstation should keep the cable length as short as reasonably possible. Large bundles of cable should be supported by

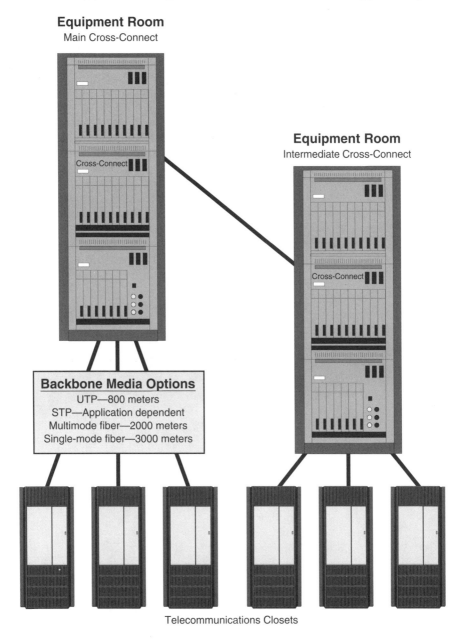

Fig. 2-8. Backbone cabling in star topology.

cable tray or conduit, the cable should maintain adequate separation from power lines and other potential sources of EMI (see EIA/TIA Standard 569, Table 10.4-1), and all cabling must be properly marked by the installer to show where each end of each cable terminates (EIA/TIA Standard 606 establishes procedures for this important aspect of your cable plant).

Chapter 3

Cable Installation

As we have already seen, different types of cable can carry signals at different speeds. There are also considerations of bandwidth, pulling characteristics, connection requirements, UL ratings, insulation ratings, shielding, low smoke ratings, and the like. These are critical factors that must be carefully considered and coordinated in the planning of any network.

During the actual installation, the various cables will be roughed in much the same as is done for common electrical cables. But as mentioned earlier, the manufacturer's instructions must be followed precisely. Also important is the use of the proper connectors and fittings. There are many different types of such fittings, and the placement of the appropriate fittings is crucial to the final equipment connections.

When pulling main runs of cables, additional conditions must be met. In general, the installation of these conductors is accomplished by the same methods as with standard electrical wires, but extra care is necessary. Because the cable is made with far less copper, the tensile strength of the cable is proportionally reduced. Sidewall pressure also carries the risk of damaging the insulation of the individual conductors, which puts the performance of the product at risk. Some general rules follow:

1. Do not exceed a pulling tension of 20% of the ultimate breaking strength of the cable (these figures are available from the cable maker).
2. Lubricate the raceway generously with a suitable pulling compound. (Check with the manufacturer for types of lubricants that are best suited to the type of cable.
3. Use pulling eyes for manhole installations.
4. For long underground runs, pull the cable both ways from a centrally located manhole to avoid splicing. Use pulling eyes on each end.
5. Do not bend, install, or rack any cable in an arc of less than 12 times the cable diameter.

Roughing and Trimming

Like power wiring, the installation of data cabling consists of two primary phases:
- roughing in the wiring,
- then trimming it later.

During the rough-in phase, the important things are that all of the cables are put into the proper places, and that they are installed carefully (not bent too tightly, pulled too hard, skinned, or otherwise damaged. At this time, it is also important to consider the routing of the cables, especially if they are unshielded. Unshielded copper cables should never be placed too close to sources of electromagnetism, such as motor windings, transformers, ballasts, or the like.

It is also important to consider fire stopping. Note where the fire barriers in the structure are, and make sure that you make proper allowances for penetrating and resealing them.

One thing in which roughing data cable differs from roughing-in power wiring is this: You must be very sure that your cables will be protected during the construction process (while you are not there). This is critical in situations where there will be a long time between your cable installation and installing the jacks. In such situations, it is up to you to protect your cables any way that will work. If you do not protect them, they may be pulled and twisted by accident. This will damage the cables, even though the damage does not show until you test them.

In other situations, such as when you have a complete raceway system to use, there may be very little time between the cable installation and the wiring of the jacks.

Also like power wiring, it is important to leave enough extra cable at each outlet point. The recommended lengths are a minimum of 3 meters in the telecommunications closet for both twisted-pair and fiber cable, one meter for fiber and 30 centimeters for twisted-pair cable at the outlet. (Notice that when you move from power wiring to data cabling, the units of measurement switch from English to metric.) Also, remember to check your specifications for requirements of extra cable.

Trimming data cabling is pretty much the same as trimming power wiring (strip the cables, install the devices and plates, etc.), except that a lot more testing is required. When trimming power wiring, we generally test by flipping a switch or hitting the outlet with a Wiggy. Either power is present, or it is not. Testing data cabling is not so simple. Remember, we need to test not only for the presence of the signal, but also for the quality of the signal.

We'll cover the requirements and methods of testing later in the text, but for now it is important to understand that you will be spending serious time testing your cables, and documenting those test results. Get used to it.

Wiring Layouts

In an earlier section, we explained *topology* - the patterns in which data networks connect computers. Topology decides how your cables will be run. In most cases, this will be a star pattern, meaning that every data outlet gets its own home run. (This is the routing for the EIA/TIA 568 standard, which almost all new computer networks follow.)

Under this system, a Category 5 (or now, perhaps level 5e) cable runs from the outlet to a communications closet. Usually, the communications closet is a telephone closet, with a little extra equipment added to it. This can make for space problems, especially in pre-existing buildings. During your estimating and planning, bear in mind that the closets my be over-crowded. (Do something about this beforehand if at all possible.)

At the wiring closet, your cable home runs will connect to a punch-down block. (The standard method of connecting communications conductors is at a multi-terminal assembly of self-stripping, crimp connections. This component is called a *punch-down block*, *66 block* or *110 block*. Punch-down block is the generic name, 66 blocks are designed especially for voice conductors, and 110 blocks are designed especially for data conductors.)

Patch panels and punch-down blocks are used to facilitate testing, additions to the system, and modifications to the *cable plant* (cabling system).

From the punch-down block, short patch cables are run to a patch panel, and from there to a *hub*. A hub is an electronic device that takes the signals from each of the cables, and puts them into a *backbone* cable, which runs between floors of the building, and connects several hubs together. You may also hear hubs referred to as *concentrators*.

The outlet itself will almost always be one or two RJ-45 jacks, mounted on a single-gang plate. (The RJ-45 is the 8-pin modular phone plug, and is nearly universally used for data networks.)

Cable Colors

While the color coding of data cabling is not as well known (or indeed, followed) as well as the color code for power conductors, color codes do exist, and should be followed. (In power wiring, things explode if you don't use the color code. In data cabling, they simply do not work.)

The cabling administration standard (EIA-606) lists the colors and functions of data cabling as:

Blue	Horizontal voice cables
Brown	Inter-building backbone
Gray	Second-level backbone
Green	Network connections & auxiliary circuits
Orange	Demarcation point, telephone cable from Central Office
Purple	First-level backbone
Red	Key-type telephone systems
Silver or White	Horizontal data cables, computer & PBX equipment
Yellow	Auxiliary, maintenance & security alarms

Separation From Sources of Interference

We mentioned earlier that unshielded data cables should not be installed near sources of electromagnetism. There is a standard that specifies these distances for structured data cabling systems. EIA/TIA-569, the cabling pathways standard, specifies the following:

Minimum Separation Distance from Power Source at 480V or less

CONDITION	<2kVA	2-5kVA	>5kVA
Unshielded power lines or electrical equipment in proximity to open or non-metal pathways	5 in.	12 in.	24 in.
Unshielded power lines or electrical equipment in proximity to grounded metal conduit pathway	2.5 in.	6 in.	12 in.
Power lines enclosed in a grounded metal conduit (or equivalent shielding) in proximity to grounded metal conduit pathway	–	6 in.	12 in.
Transformers & electric motors	40"	40"	40"
Fluorescent lighting	12"	12"	12"

Minimum Bending Radii

According to a draft version of EIA-568, the minimum bend radius for UTP is 4 times outside cable diameter, or about one inch. For multi-pair cables the minimum bending radius is 10 x outside diameter. The minimum bend radii for Type 1A Shielded Twisted Pair (100 Mb/s STP) is 7.5 cm (3-in) for non-plenum cable, 15 cm (6-in) for the stiffer plenum-rated kind.

For optical cables not under tension, the minimum bend radius is 10 times diameter; and for cables under tension, no less than 20 times cable diameter. The standard goes on to state that no optical cable will be bent on a radius less than 3.0 cm (1.18-in).

A different standard, ISO DIS 11801 (essentially a parallel standard to the one mentioned above), for 100 ohm and 120 ohm balanced cable lists three different minimum bend radii. Minimum for pulling during installation is 8 times cable diameter, minimum installed radius is 6 times for riser cable, and 4 times cable diameter for horizontal runs. For fiber optic cables, the requirements are the same as those stated above.

Some manufacturers recommendations differ from the above, so it is worth checking the spec sheet for the cable you plan to use.

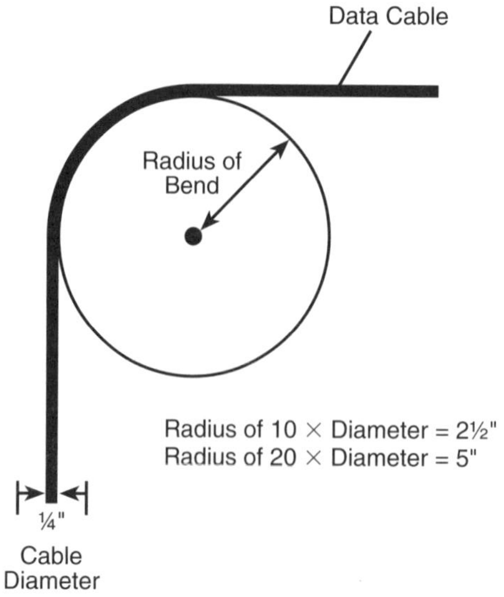

Fig. 3-1. Minimum bending Radii.

Installation Requirements

Article 800 of the NEC covers communication circuits, such as telephone systems, computer networks built around telephone-style cables, and outside wiring for fire and burglar alarm systems. Generally these circuits must be separated from power circuits and grounded. In addition, all such circuits that run out of doors (even if only partially) must be provided with circuit protectors (surge or voltage suppressers).

Article 725 of the National Electrical Code covers a few types of network cabling. Most Category 5 cables, however, are rated under article 800, rather than under article 725.

The requirements of article 800 are these:

Conductors Entering Buildings

If communications and power conductors are supported by the same pole, or run parallel in span, the following conditions must be met:
1. Wherever possible, communications conductors should be located below power conductors.
2. Communications conductors cannot be connected to cross arms.
3. Power service drops must be separated from communications service drops by at least 12 inches.

Above roofs, communications conductors must have the following clearances:
1. Flat roofs: 8 feet.
2. Garages and other auxiliary buildings: None required.
3. Overhangs, where no more than 4 feet of communications cable will run over the area: 18 inches.
4. Where the roof slope is 4 inches of rise for every 12 inches horizontally: 3 feet.

Underground communications conductors must be separated from power conductors in manhole or handholes by brick, concrete, or tile partitions.

Communications conductors should be kept at least 6 feet away from lightning protection system conductors.

Circuit Protection

Protectors are surge arresters designed for the specific requirements of communications circuits. They are required for all aerial circuits not confined with a *block*. (Block here means city block.) They must be installed on all circuits with a block that could accidentally contact power circuits over 300 volts to ground. They must also be listed for the type of installation.

Other requirements are the following:

Metal sheaths of any communications cables must be grounded or interrupted with an insulating joint as close as practicable to the point where they enter any building (such point of entrance being the place where the communications cable emerges through an exterior wall or concrete floor slab, or from a grounded rigid or intermediate metal conduit).

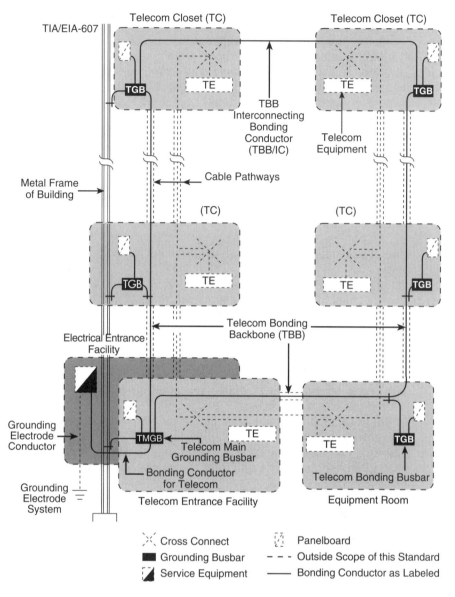

Fig. 3-2. Schematic of a typical datacom grounding system.

Grounding conductors for communications circuits must be copper or some other corrosion-resistant material, and have insulation suitable for the area in which it is installed.

Communications grounding conductors may be no smaller than No. 14.

The grounding conductor must be run as directly as possible to the grounding electrode, and be protected if necessary.

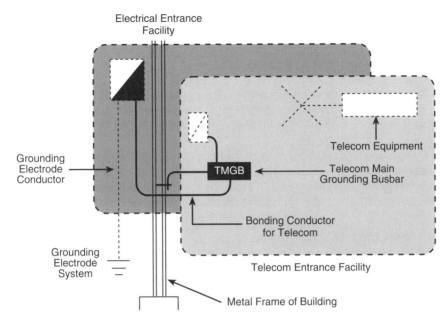

Fig. 3-3. Telecom main grounding busbar (TMGB).

If the grounding conductor is protected by metal raceway, it must be bonded to the grounding conductor on both ends.

Grounding electrodes for communications ground may be any of the following:

1. The grounding electrode of an electrical power system.
2. A grounded interior metal piping system. (Avoid gas piping systems for obvious reasons.)
3. Metal power service raceway.
4. Power service equipment enclosures.
5. A separate grounding electrode, bonded to the power system grounding electrode.

If the building being served has no grounding electrode system, the following can be used as a grounding electrode:

1. Any acceptable power system grounding electrode. (See Section 250-81.)
2. A grounded metal structure.

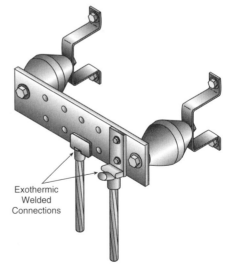

Fig. 3-4. Exothermic welded connections on a TMGB.

39

3. A ground rod or pipe at least 5 feet long and ½ inch in diameter. This rod should be driven into damp (if possible) earth, and kept separate from any lightning protection system grounds or conductors.

Connections to grounding electrodes must be made with approved means.

As mentioned above, if the power and communications systems use separate grounding electrodes, they must be bonded together with a No. 6 copper conductor. Other electrodes may be bonded also. This is not required for mobile homes.

For mobile homes, if there is no service equipment or disconnect within 30 feet of the mobile home wall, the communications circuit must have its own grounding electrode. In this case, or if the mobile home is connected with cord and plug, the communications circuit protector must be bonded to the mobile home frame or grounding terminal with a copper conductor no smaller than No. 12.

Interior Communications Conductors

Communications conductors must be kept at least 2 inches away from power or Class 1 conductors, unless they are permanently separated from them or unless the power or Class 1 conductors are enclosed in one of the following:
1. Raceway.
2. Type AC, MC, UF, NM, or NM cable, or metal-sheathed cable.

Communications cables are allowed in the same raceway, box, or cable with any of the following:
1. Class 2 and 3 remote-control, signaling, and power-limited circuits.
2. Power-limited fire protective signaling systems.
3. Conductive or nonconductive optical fiber cables.
4. Community antenna television and radio distribution systems.

Communications conductors are not allowed to be in the same raceway or fitting with power or Class 1 circuits.

Communications conductors are not allowed to be supported by raceways unless the raceway runs directly to the piece of equipment the communications circuit serves.

Openings through fire-resistant floors, walls, etc. must be sealed with an appropriate firestopping material.

Any communications cables used in plenums or environmental air-handling spaces must be listed for such use.

Requirements of Article 725

Article 725 of the National Electric Code, or NEC, covers Class 1, 2, and 3 remote control, signaling, and power-limited circuits. This article can be very confusing if you do not understand that circuits designated as Class 1, 2, or 3 can be either *power-limited* circuits, OR *signaling* circuits.

Every circuit covered by this article is Class 1, 2, or 3.

Some of them are power-limited, and *some* are signaling circuits.

It is important to note that this article does not apply to such circuits that are part of a device or appliance. It applies only to separately installed circuits.

Definitions

One of the more difficult parts of this article is that it involves a lot of terms with which most of us are not familiar. The key terms are as follows:

Class 1: Circuits that are supplied by a source that has an output of no more than 30 volts (AC or DC) and 1000 volt-amperes. (*Volt-amperes* is essentially the same thing as *Watts*. You will find the term volt-amps used when relating to transformers and similar devices, since it is technically more correct for such uses than Watts.) See section 725-11.

Class 2: Circuits that are inherently limited in capacity. They may require no overcurrent protection, or may have their capacity regulated by a combination of overcurrent protection and their power source. There are a number of voltage, current, and other characteristics that define class 2 circuits. These characteristics are detailed in tables 725-31(a) and 725-31(b).

Class 3: Class 3 circuits are very similar to class 2 circuits. They are inherently limited in capacity. They may require no overcurrent protection, or may have their capacity regulated by a combination of overcurrent protection and their power source. The voltage, current, and other characteristics that define class 3 circuits and differentiate them from 2 circuits are detailed in tables 725-31(a) and 725-31(b).

Power-Limited: This refers to a circuit that has a self-limited power level. That is, whether because of impedance, overcurrent protection, or other power source limitations, these circuits can only operate up to a limited level of power.

Remote control and signaling circuits: These are class 1, 2, or 3 circuits that do not have a limited power rating.

Supply side: This is essentially the same thing as "line side". This term is used instead of "line" because of possible confusion. (Someone might think it refers to the power circuits, rather than simply to the power that feeds the circuit.)

CL2, CL3P, CL2R, etc.: These are specific types of power-limited cables. See table 725-50.

PLTC: Power Limited Tray Cable.

Requirements

A remote control, signaling, or power limited circuit is the part of the wiring system between the load side of the overcurrent device or limited power supply and any equipment connected to the circuit.

There are a lot of different requirements listed in article 725. The ones we will cover here will be the most important of the installation requirements. We will begin with class 1 requirements. The requirements are as follows:

Except for transformers, the power supplies that feed class 1 circuits must be protected by an overcurrent device that is rated no more than

167% of the power supply's rated current. This overcurrent device can be built into the power supply; but if so, it can not be interchangeable with devices of a higher rating. In other words, you cannot use any interchangeable fuse (as most are) in this power supply. Transformers feeding these circuits have no requirements except those stated in article 450.

All remote control circuits that could cause a fire hazard must be classified as class 1, even if their other characteristics would classify them as Class 2 or class 3. This applies also to safety control equipment. (This does not include heating and ventilating equipment.

Class 1 circuits are not allowed in the same cable with communications circuits.

Power sources for these circuits must have a maximum output (note: this is *maximum output*, not *rated power*; the two terms are very different) of no more than 2,500 VA. This does not apply to transformers.

Class 1 remote control and signaling circuits can operate at up to 600 volts, and their power sources need not be limited.

All conductors in class 1 circuits that are #14 AWG or larger must have overcurrent protection. Derating factors cannot be allowed. #18 conductors must be protected at 7 amps or less, and #16 conductors at 10 amps or less. There are three exceptions to this. See 725-12.

Any required overcurrent devices must be located at the point of supply. See 725-13 for two exceptions.

Class 1 circuits of different sources can share the same raceway or cable, provided that they all have insulation rated as high as the highest voltage present.

If only class 1 conductors are in a raceway, the allowable number of conductors can be no more than easy installation and heat dissipation will allow.

Definitions of Class 2 and Class 3 circuits:

These are circuits that are inherently limited in capacity. They may require no overcurrent protection, or may have their capacity regulated by a combination of overcurrent protection and their power source. There are a number of voltage, current, and other characteristics that define these circuits. These characteristics are detailed in tables 725-31(a) and 725-31(b).

If you refer to the tables referenced above, you will find that there are quite a few combinations of circuit characteristics that can make the circuit Class 2 or Class 3. Note that the characteristics for AC and DC circuits are different. These tables each have two particular groupings - Circuits that require overcurrent protection, and circuits that do not.

The only substantial differences between Class 2 and Class 3 circuits is that Class 3 circuits generally have higher voltage and power ratings than Class 2 circuits.

Requirements

The basic requirements for Class 2 and Class 3 circuits are as follow:
Power supplies for Class 2 or 3 circuits may not be connected in

parallel unless specifically designed for such use.

The power supplies (when necessary) that feed class 2 or 3 circuits must be protected by an overcurrent device that can not be interchangeable with devices of a higher rating. You cannot use an interchangeable fuse in this power supply. The overcurrent device can be built in to the power supply.

All overcurrent devices must be installed at the point of supply.

Transformers that are supplied by power circuits may not be rated more than 20 amps. They may, however, have #18 AWG leads, so long as the leads are no more than 12 inches long.

Class 2 or 3 conductors must be separated at least 2 inches from all other conductors, except in the following situations:
1. If the other conductors are enclosed in raceway, metal-sheathed cables, metal-clad cables, NM cables, or UF cables.
2. If the conductors are separated by a fixed insulator, such as a porcelain or plastic tube.

Class 2 or 3 conductors may not be installed in the same raceway, cable, enclosure, or cable tray with other conductors, except:
1. If they are separated by a barrier.
2. In enclosures, if Class 1 conductors enter only to connect equipment which is also connected to the Class 2 and/or 3 circuits.
3. In manholes, if the power or Class 1 conductors are in UF or metal-enclosed cable.
4. In manholes, if the conductors are separated by a fixed insulator, in addition to the insulation of the conductors.
5. In manholes, if the conductors are mounted firmly on racks.
6. If the conductors are part of a hybrid cable for a closed loop system. (See article 780.)

When installed in hoistways, Class 2 or 3 conductors must be enclosed in rigid metal conduit, rigid nonmetallic conduit, IMC, or EMT. They may be installed in elevators as allowed by section 620-21.

In shafts, Class 2 or 3 conductors must be kept at least 2 inches away from other conductors.

Any cables that are used for Class 2 or 3 systems must be marked as resistant to the spread of flame.

All Class 2 and 3 cables are marked, listing the areas in which they may be installed. The cables may be installed in the listed areas only.

Two or more Class 2 circuits can be installed in the same enclosure, cable, or raceway, so long as they are all insulated for the highest voltage present.

Two or more Class 3 circuits may share the same raceway, cable, or enclosure.

Class 2 and Class 3 circuits can be installed in the same raceway or enclosure with other circuits, provided the other circuits are in a cable of one of the following types:
- Power-limited signaling cables (see article 760)
- Optical fiber cables

- Communication cables (see article 800)
- Community antenna cables (see article 820)

When Class 2 or 3 conductors extend out of a building, and are subject to accidental contact with systems operating at over 300 volts to ground (not 300 volts between conductors - but 300 volts to ground), they must meet all the requirements of section 800-30 for communication circuits.

Cabling Classifications

Following are the cable classifications which are applicable to data cabling are these:

Article 725, Class 2:

725-38(b)1	CL2X	Class 2 cable, limited use
725-38(b)1	CL2	Class 2 cable
725-38(b)2	CL2R	Class 2 riser cable
725-38(b)3	CL2P	Class 2 plenum cable

Article 800:

800-3(b)1	CMX	Communications cable limited use
800-3(b)1	CM	Communications cable
800-3(b)2	CMR	Communications riser cable
800-3(b)3	CMP	Communications plenum cable

Article 770:

OFNP (Optical Fiber Nonconductive Plenum)

OFNR (Optical Fiber Nonconductive Riser)

EIA/TIA-568

By far the most commonly used standard for structured wiring systems is EIA/TIA 568, or 568A. The EIA/TIA-568 wiring standard recognizes four cable types and two types of telecommunications outlets.
- Some of the parameters of 568 are the following:
- Up to 50,000 users
- Facilities of up to 10 million square feet
- 90 meter horizontal distance limit between closet and desktop
- 4 pairs of conductors to each outlet - all must be terminated
- 25-pair cables may *not* be used (crosstalk problems)
- May not use old wiring already in place
- Bridge taps and standard telephone wiring schemes may not be used
- Requires careful installation procedures
- Requires extensive testing procedures

Following are the standard materials used for a structured cabling (568) system:

Four-pair 100 ohm UTP cables. The cable consists of № 24 AWG thermoplastic insulated conductors formed into four individually-twisted pairs and enclosed by a thermoplastic jacket. Four-pair, № 22 AWG

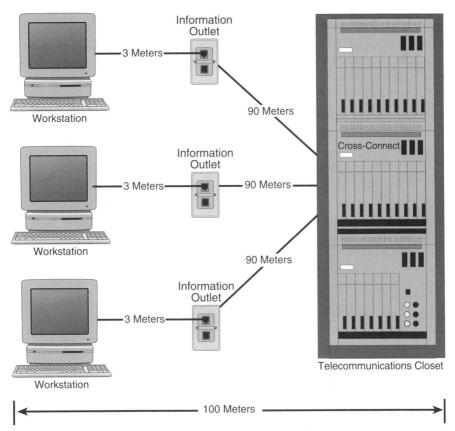

Fig. 3-5. Maximum distances for horizontal cabling. Shown are the standard distance limitations per the structured cabling standard.

cables that meet the transmission requirements may also be used. Four-pair, *shielded* twisted pair cables that meet the transmission requirements may also be used.

The pair twists of any pair shall not be exactly the same as any other pair. The pair twist lengths shall be selected by the manufacturer to assure compliance with the crosstalk requirements of this standard.

Color Codes

Pair 1   White-Blue   (W-BL)   Blue (BL)
Pair 2   White-Orange (W-O)   Orange (O)
Pair 3   White-Green (W-G)   Green (G)
Pair 4   White-Brown (W-BR)   Brown (BR)

Cable Specifications

The diameter of the completed cable shall be less than 6.35mm (0.25 in).
The ultimate breaking strength of the completed cable is 90 lb minimum.
Maximum *pulling* tension should not exceed 25 lb to avoid stretching.

The cable tested shall withstand a bend radius of 25.4 mm (1 in) at a temperature of -20 C without jacket or insulation cracking

The resistance of any conductor shall not exceed 28.6 ohms per 305 m (1000 ft) at or corrected to a temperature of 20 C.

The resistance unbalance between the two conductors of any pair shall not exceed 5% when measured at or corrected to a temperature of 20 C.

The mutual capacitance of any pair at 1 kHz shall not exceed 20 nF per 305 M (1000 ft).

The mutual capacitance of any pair at 1 kHz and measured at or corrected at a temperature of 20C, shall not exceed 17 nF per 305 m (1000 ft) for category 4 and category 5 cables.

The capacitance unbalance to ground at 1 kHz of any pair shall not exceed 1000 pF per 305 m (1000 ft).

The attenuation of any pair shall not exceed the following values:

Frequency (MHZ)	Max Attenuation (db/1000')
0.064	2.8
0.256	4.0
0.512	5.6
0.772	6.8
1.0	7.8
4.0	17
8.0	26
10.0	30
16.0	40

The maximum attenuation of any pair shall be less than the following, in the frequency range for 0.772 MHZ to the highest referenced frequency.

Maximum Attenuation dB per 1000 ft @ 20C

Frequency MHZ	Category 3	Category 4	Category 5
0.064	2.8	2.3	2.2
0.256	4.0	3.4	3.2
0.512	5.6	4.6	4.5
0.772	6.8	5.7	5.5
1.0	7.8	6.5	6.3
4.0	17	13	13
8.0	26	19	18
10.0	30	22	20
16.0	40	27	25
20.0	–	31	28
25.0	–	–	32
31.25	–	–	36
62.5	–	–	52
100	–	–	67

The NEXT coupling loss between any two pairs within a cable shall be equal to or greater than the following values:

Frequency (MHZ)	NEXT Loss Worst Pair (db @ 1000 ft)
0.15	54
0.772	43
1.0	41
4.0	32
8.0	28
10.0	26
16.0	23

Following are the values of *worst pair* NEXT loss at specific frequencies:

Frequency MHZ	Category 3	Category 4	Category 5
0.150	54	68	74
0.772	43	58	64
1.000	41	56	62
4.000	32	47	53
8.000	28	42	48
10.000	26	41	47
16.000	23	38	44
20.000	–	36	42
25.000	–	–	41
31.250	–	–	40
62.500	–	–	35
100.000	–	–	32

The NEXT loss of patch cables shall be equivalent to the NEXT loss of equal length of horizontal cables over the range of frequencies specified for any given cable category. The NEXT Loss for connectors should be better than the NEXT values for any given cable category specified above.

Telecommunications Outlet Specification

100-ohm UTP Cable - Each four-pair cable shall be terminated in an eight-position modular jack in the work area. The 100-ohm UTP telecommunications outlet shall meet the requirements described in EIA/TIA-570

150 ohm STP Cable - The telecommunications connector used for terminating the 150-ohm STP cable shall be that specified by ANSI/IEEE 802.5 for the

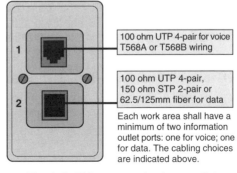

Fig. 3-6. Telecommunications outlet.

media interface connector. This connector shall be designed so that like units will mate when oriented 180 degrees with respect to each other.

Category 5 Cabling

Category 5 cable is used for high-speed data transmission, and is increasingly being used for other communications uses. It is rated up to 100 MHZ and is intended for data rates up to 100 Mbps. Category 5 specifications are designed to minimize the effects of the connecting hardware on the cabling system.

But to accomplish high-speed transmission, category 5 cabling requires very tight design characteristics. In particular, the conductors in category 5 cables must be very tightly twisted. Furthermore, this tight twist pattern can not be altered during installation. Even at terminations, no more than 13 millimeters (approximately one half inch) of conductor can be unwound. While working with a large number of cables terminating at one block, this can be difficult.

The minimum bending radius for category 5 cables is typically four times the cable diameter. This is not especially difficult for installers, although it can be difficult in tight areas. When being pulled into place, no more than 25 pounds of tension can be applied to the cable. One easily missed hazard of category 5 cabling is that the use of *tie-wraps* can damage the cable's performance. When tie-wraps are cinched down tightly on these cables, they deform the pattern of the twists, and can permanently damage the electronic characteristics of the cable badly enough that it will not handle high frequency signals.

Category 5 compliance ensures application-independence. It defines transmission performance for all pair combinations for both the RJ45 connector interface (mated plug and jack) and the associated cable. Finally, Category 5 provides a platform for implementing new high performance LAN technology regardless of the pair combination required.

Standard Networking Configurations

Here are the standard configurations for several common networks:
- ATM 155 Mbps uses pairs 2 and 4 (pins 1-2, 7-8)
- Ethernet 10Base-T uses pairs 2 and 3 (pins 1-2, 3-6)
- Ethernet 100Base-T4 uses pairs 2 and 3 (pins 1-2, 3-6)
- Ethernet 100Base-T8 uses pairs 1,2,3 and 4 (pins 4-5, 1-2, 3-6, 7-8)
- Token-Ring uses pairs 1 and 3 (pins 4-5, 3-6)
- 100VG-AnyLAN uses pairs 1,2,3 and 4 (pins 4-5, 1-2, 3-6, 7-8)

Ethernet 10BASE-T Cabling and Patch Cords

Ethernet 10Base-T Straight Thru patch cord:

RJ45 Plug		RJ45 Plug	
T2 1	White/Orange............	1 TxData +	pair2 \—
R2 2	Orange	2 TxData -	
T3 3	White/Green..............	3 RecvData +	

R1 4   Blue 4 pair3
T1 5   White/Blue 5
R3 6   Green .... 6 RecvData -
T4 7   White/Brown 7
R4 8   Brown 8

Ethernet 10Base-T Crossover patch cord:

This cable is used to cascade hubs, or for connecting two Ethernet stations back-to-back without a hub. Note pin numbering above.

RJ45 Plug   **RJ45 Plug**

1 Tx+ .... Rx+ 3
2 Tx- .... Rx- 6
3 Rx+ .... Tx+ 1
6 Rx- .... Tx- 2

A simple way to make a crossover patch cable is to take a dual-jack surface mount box and make the crossover between the two jacks. This allows using standard patch cables, and avoids the nuisance of having a crossover cable find its way into use in place of a regular patch cable.

The color code used in stranded patch cables (which are used for a few special systems) is different from solid-conductor cables. For NorTel Digital Patch Cable (DPC), the coding is:

Pair 1: Green & Red
Pair 2: Yellow & Black
Pair 3: Blue & Orange
Pair 4: Brown & Gray

Chapter 4

Optical Fiber

Some of the more common terms used to describe optical fiber are a "light tube" or a "conduit for light". These terms have led a lot of people to think that there is some type of hole in the middle of an optical fiber - this is not the case.

Nonetheless, optical fiber does function as a "tube" or "conduit" for light. Light does flow down the center of a fiber like water flows through a pipe. As you can see in **Fig. 4-1**, there are three concentric layers to an optical fiber. Light flows only through the glass *core* of the fiber. The *cladding* (which is a different type of glass) serves as a barrier to keep the light within the core, functioning much like a mirrored surface. The *buffer* (also called *coating*) layer has nothing to do with light transmission, and is used only for mechanical strength and protection.

Light is kept to the core of the fiber, and flows though the core as water would flow through a tube. We could even say that the fiber is a "virtual" tube. The light stays in the center of the fiber, not because there is a physical opening there, but because the cladding glass reflects any escaping light back to the core.

These "light tubes" are very thin strands of ultra-pure glass. The dimensions of typical fibers are as follow:

Core   8 - 62.5 microns (a micron is one millionth of a meter)
Cladding   125 microns
Coating   250 microns

The dimensions shown above are diameters. The core is one density of

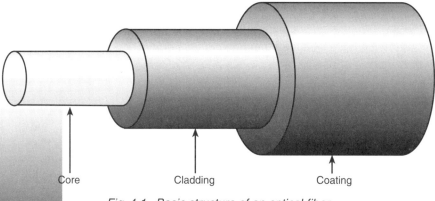

Fig. 4-1. Basic structure of an optical fiber.

glass, the cladding is a second grade of glass, and the coating is plastic.

Fibers

The two main types of optical fibers that are commonly used today are the following:
1. **Single-mode fibers.** A single-mode fiber allows only one light wave ray to be transmitted down the core. The core is extremely small, usually between 8 and 9 microns. Because of quantum mechanical effects, the light traveling in the very narrow core stays together in packets, rather than bouncing around the ore of the fiber. Thus single-mode fiber has an advantage over all other types in that it can handle far more signal, and over far greater distances.
2. **Multi-mode, *graded-index* fibers.** Graded-index fibers contain many layers of glass, each with a lower index of refraction as it moves outward from the center. Since light travels faster in the glass with lower indexes of refraction, the light waves refracted to the outside of the fiber are speeded up to match those traveling in the center. The result that this type of fiber allows for high speed data to be transmitted over a reasonably long distance. Multi-mode fibers are used with LED light sources, which are less expensive than the laser light sources that are used for single-mode. Graded index fibers come in core diameters of 50, 62.5, 85, and 100 microns.

Fiber Sizing

The size of an optical fiber is referred to by the outer diameter of its core and cladding. For example, a size given as 62.5/125 indicates a fiber that has a core of 50 microns and a cladding of 125 microns. The coating is not typically mentioned in the size, because it has no effect on the light-carrying characteristics of the fiber.

The *core* is the part of the fiber that actually carries the light pulses that are used for transferring data. This core may be made of either plastic or glass. The size of the core is important, as core sizes of joined fibers must match. Larger cores have greater light-carrying capacity than smaller cores, but may cause greater signal distortion.

The *cladding* sets a boundary around the fiber, so that light running into this boundary is reflected back into the cable. This keeps the light moving down the cable, keeping it from escaping. Claddings can be made of either glass of plastic, and always have a different density than that of the core. (If they did not have a different density [index of refraction], they could not reflect escaping light back into the core.)

Coatings are typically multiple layers of ultraviolet curable acrylate plastic. This is necessary to add strength to the fiber, to protect it, and to absorb shock. These coatings come between 25 and 100 microns thick. (One micron is equal to one millionth of a meter. For comparison purposes, a sheet of paper has a thickness of approximately 25 microns.) Coatings can be stripped from the fiber (and must be for terminating), either mechanically or chemically, depending what type of plastic is used.

Fiber Type	Core/Cladding Diameter(m)	Attenuation (dB-km)			Bandwidth (MHz-km)
		850 nm	1300 nm	1550nm	
Step Index	200/240	6			50
Multimode / Graded Index	50/125	3	1		600
	62.5/125	3	1		500
	85/125	3	1		500
	100/140	3	1		300
Singlemode	8-9/125		0.5	0.3	high
Plastic	1 mm	(1 dB/m @ 665 nm)			Low

Table 4-1. Fiber types and typical specifications.

Cabling

To suitably protect our glass fibers, we package them in cabling. It is a misconception that fiber cables are fragile. Many people believe that if you drop a fiber cable on a hard surface, it will shatter - "After all, it's made of glass!" This is not so.

Actual fiber itself (not the *cable*, but only the thin *fiber*) is fairly fragile, although it is surprisingly flexible, and will not break easily like sheet glass. (Actually, it is several times as strong as steel; but, being so thin, it can be broken fairly easily.)

Fiber cables, however, are not at all fragile. In fact they are often more durable than copper communication cables. Optical cables encase the glass fibers in several layers of protection, as is shown in figure X.

The first protective layer is the coating that we mentioned earlier. The next layer of protection is a *buffer* layer. The buffer is typically extruded over the coating to further increase the strength of the single fibers. This buffer can be of either a *loose tube* or *tight tube* design. Most datacomm cables are made using either one of these two constructions. A third type, the ribbon cable, is frequently used in the telecommunications work, and may be used for datacomm applications in the future; it uses a modified type of tight buffering.

After the buffer layer, the cable contains a *strength member.* Most commonly, the strength member is Kevlar® fabric, the material that bullet-

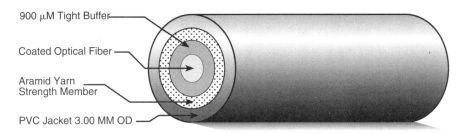

Fig. 4-2. This is a cross section of a tight-buffered fiber cable having one buffered fiber about 900 microns in diameter.

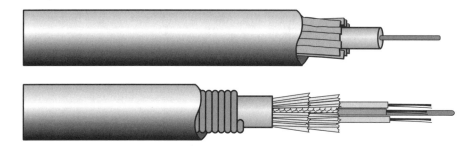

Fig. 4-3. The top cable is a breakout cable. The one on the bottom is an armored loose tube cable.

proof vests are made from. The strength member not only protects the fiber, but is used to carry the tensions of pulling the cable. (You can never ever pull the fibers themselves.)

After the strength member finally comes the outer jacket of the cable, which is typically some type of polyethylene or PVC. In many cases, however, there will be additional *stiffening members* which also increase the cable's strength and durability.

Fiber Connectors

As the fiber optic field began to develop, one of the biggest mechanical problems that existed was permanently fixing the fibers at their ends. Since they have such a small diameter, they must be held rigidly in place and accurately aligned to mate with other fibers, light sources, or light detectors. The first fiber connectors were difficult to install. They used a variety of glues, ovens, and long, difficult polishing methods.

Between then and now, however, almost everything has changed. While terminating a fiber is not yet as easy as installing a coaxial cable connector, it has become far, far easier. Now, terminating fibers (termination involves installing a connector and polishing the face of the fiber) can be done in half the previous time. And the process continues to get easier as time goes on. Within a few years it should be quite simple.

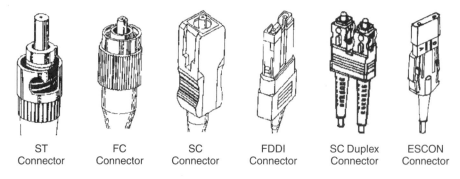

ST Connector   FC Connector   SC Connector   FDDI Connector   SC Duplex Connector   ESCON Connector

Fig. 4-4. Connectors.

Splices

While fiber connectors are used to make non-permanent connections at the ends of fibers, splices are used to join two ends of fiber to each other permanently. There are two primary ways this is done: by *fusion*, that is, by melting the pieces of glass together; and by *mechanical* means. The critical factors in splicing are the following:

1. That the fiber joint passes light without loss.
2. That the joint is mechanically secure; (ie. that it will not be easily broken).

Datacommunications	Telecommunications
(mostly multimode)	(mostly single-mode)
SMA (decreasing)	FC/PC (widely used)
ST (most commonly used)	ST (single-mode version)
SC (specified for newer systems)	SC (growing)
FDDI (duplex)	D4 (decreasing)
ESCON (duplex)	Biconic (decreasing)

Table 4-2. Common applications for optical connectors.

Testing

When installing a fiber system (the whole system of optical fibers is often called a *cable plant*), it must be tested to assure that it will properly perform. The purpose of testing is to make sure that light will pass through the system properly. The main types of optical testing used in the field are the following:

Continuity testing. This is a simple visible light test. Its purpose is to make sure that the fibers in the cables are continuous, that is, that they are not broken. This is done with a modified type of flashlight device and the naked eye, and takes only a few minutes to perform.

Power testing. This is to accurately measure the quality of optical fiber links. A calibrated light source puts infrared light into one end of the fiber, and a calibrated meter measures the light arriving at the other end of the fiber. The loss of light in the fiber is measured in decibels.

OTDR testing. The OTDR is a piece of equipment properly called an *Optical Time Domain Reflectometer*. This device uses light backscattering to analyze fibers. In essence, the OTDR takes a snapshot of the fiber's optical characteristics. It sends a high powered pulse into the fiber and measures the light scattered back toward the instrument. The OTDR can be used to locate fiber breaks, splices and connectors, as well as to measure loss. However, the OTDR method of loss measurement may not give the same value for loss as a source and power meter, duie to the different methods of measuring loss. In addition, the OTDR gives a graphic display of the status of the fiber being tested. Another advantage is that it requires access to only one end of the fiber.

As useful as the OTDR is, however, it is not necessary in the majority of situations. A power meter and source are used to test the loss of fiber optic cable, simulating the way the fibers are used, and measuring the light lost from one end of the cable to the other. In addition, OTDRs are quite expensive. Even when they are necessary, many installers prefer to

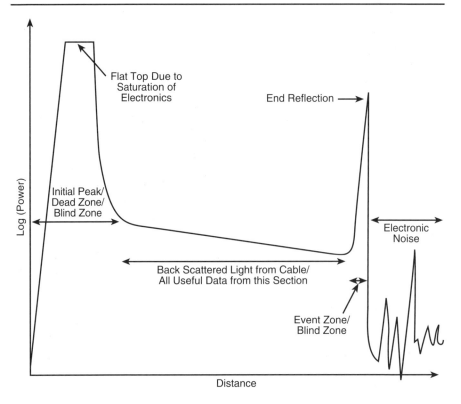

Fig. 4-5. Basic features of the OTDR trace.

rent them, rather than purchasing them.

Other Basic Concepts

In addition to the things we have so far covered, there are several other basic concepts that you must understand:

Attenuation. Attenuation is the weakening of an optical signal as it passes through a fiber. Attenuation is signal loss. Attenuation in an optical fiber is a result of two factors, *absorption* and *scattering*.

Absorption is caused by the absorption of the light and conversion to heat by molecules in the glass. Primary absorbers are residual deposits of chemicals that are used in the manufacturing process to modify the characteristics of the glass. This absorption occurs at definite *wavelengths* (the wavelength of light signifies its color and its place in the electromagnetic spectrum). This absorption is determined by the elements in the glass at wavelengths around 1000 nm (nanometers), 1400 nm and above 1600 nm.

The largest cause of attenuation is *scattering*. Scattering occurs when light collides with individual atoms in the glass.

Fiber optic systems transmit in the "windows" created between the absorption bands at 850 nm, 1300 nm and 1550 nm wavelengths, for which lasers and detectors can be easily made.

Networks. To communicate between several pieces of equipment (for example, between 20 different computers in an office), they must all be connected together. To do this, we must do two things:
1. Develop a logical method of connecting them. Should we tie them all to a central point? Should we connect them in a loop?
2. Provide a definite protocol for communicating. If the machines do not "talk to each other" in some type of order, the whole system will collapse in a jumble of signals that the machines cannot separate or interpret.

Many types of networks have been developed, each with its own strengths and weaknesses. The many network names you have probably heard, such as Ethernet, 10base T, FDDI, ATM, and Token Ring are simply different methods of connecting computers together.

Bandwidth. Bandwidth is the range of signal frequencies or bit rate at which a fiber system can operate. In essence, it is a measure of the amount of signal that can be put through a fiber. Higher bandwidth means more data per second; lower bandwidth means less signal.

Dispersion. There are two potentially confusing terms that you will come across in your readings: *Chromatic dispersion* and *modal dispersion*. In both of these terms, *dispersion* refers the spreading of light pulses, until they overlap one another, and the data signal is distorted and lost. *Chromatic* refers to color, and *modal* primarily refers to the light's path. Thus we can state in simple terms that:

Chromatic dispersion = Signal distortion due to Color.

Modal dispersion = Signal distortion due to Path.

Note that dispersion is NOT a loss of light, it is a distortion of the signal. Thus dispersion and attenuation are two different and unrelated problems: Attenuation is a loss of light; dispersion is a distortion of the light signals.

Installation

Some of the basic installation requirements are the following:
- When pulling optical fibers, the stresses must be places on the cable's strength member, never on the fibers themselves.
- The cables may not be tightly bent or kinked.
- At least one meter of extra cable should be left at every outlet point.

Terminating optical fibers is the most time-consuming part of the process. There are two primary concerns:
1. That the terminations are well-made. That is, that they pass light correctly, exhibiting low db losses, and that they are physically durable.
2. That safety precautions are observed. Broken pieces of bare fiber can act like tiny needles, and since they are both small and clear, they can be quite a problem.

Saftey Precautions

Live optical fiber ends (*live* fibers are those with signals being sent through them) should not be inspected by technicians. All fibers should be *dark* (no signal being transmitted) when inspected. This must be done carefully, since the light used in the majority of optical systems is not visible to the human eye.

If there is a risk of fibers being inspected live, especially when the system light source is a laser, all technicians working on the system should wear protective glasses which have infrared filtering.

Fiber optic work areas must be clean, organized, well-lit, and equipped with a bottle or other suitable container for broken or stray fiber pieces.

No food, drink, or smoking should be allowed in areas where fiber optic cables are spliced or terminated, or in any area where bare fibers are being handled.

Technicians making fiber terminations or splices or working with bare fiber need double-sided tape, or some other effective means, for picking up broken or stray pieces of fiber. Their work areas will have to be repeatedly and consistently cleared of all bare fiber pieces. All bare fiber pieces must be disposed of so that they cannot escape and cause a hazard. (For example, bare fibers should be sealed in some type of bottle or container before being dumped into a wastebasket.)

Finally, all technicians working on bare fiber should thoroughly wash their hands immediately when leaving the work area, check their clothing, and pat themselves with clean tape to remove any stray pieces of bare fiber.

The Fiber Loss Budget

The central concern with optical fiber testing is that the system has enough light making its way through the fibers to operate the equipment correctly. When fiber systems are designed they are given a *loss budget*, which is based on the electronic equipment being used. Almost all of our field testing is done to assure that this loss budget will not be violated. When performing cable installations and terminations, it is important to have a copy of the system's loss budget, so you know how much loss you can afford.

Although there is seldom a problem of exceeding a 12-16 dB loss budget, bear in mind that the requirements of your contract may be more stringent. Nonetheless, without the loss budget, you don't really know what the system's real parameters are.

Design Short-Cuts

Following are several design short-cuts. These are generally the most applicable and cost-effective design choices in an open marketplace. Note, however, that these are general recommendations, and that situations may dictate otherwise.

Fiber Choice

Multimode - Use 62.5/125 fiber. 62.5/125 is the de facto standard. It offers higher launched power than the 50/125 fiber at a modest price increase. In addition, most standards for data communication are focused the 62.5/125 fiber. There is no indication that this will change any time soon.

Singlemode - Use 1300 nm singlemode fiber. Systems designed to operate at this wavelength have lower cost than do 1550 nm systems. Do not choose fiber designed for both 1300 and 1550 nm unless you expect to use wavelength division multiplexing or optical amplifiers in the future.

Cable Design

Indoor - For short distances [<1200-1335 feet], use break-out cables. For longer distances, use premise (distribution) cables. If your environment is especially rugged, use break-out design rather than premise cable. The price premium is insurance against future maintenance cost.

Use all-dielectric design. These cables are not subject to grounding requirements.

If plenum cables required, look for plenum-rated PVC products. Teflon plenum cables are not only more expensive, but they are often hard to find or will be available only after a long delay.

Outdoor cables - Use water-blocked and gel-filled, loose tube designs.

Consider ribbon designs (also water-blocked and gel-filled) if the cable will have 36 or more fibers. Ribbon-based cables are cheaper *per fiber* than other types of cables.

If mid-span access is important, use the stranded loose-tube design.

Use all-dielectric design. For the reason specified earlier - no grounding requirements.

Indoor/Outdoor Cables - If cable must be installed both indoors and outdoors, use indoor/outdoor rated cable. In so doing, you can eliminate a splice or connector pair. This design has an easily removable outdoor jacket over an inner structure that meets NEC requirements.

Fiber Performance

Multimode - Choose dual wavelength specifications.

Wavelength	850/1300 nm
Attenuation rate	3.75/1.0 dB/ km
Bandwidth-distance product	160/500 MHZ-km
NA (numerical aperture)	.275 nominal

Singlemode - Choose single wavelength specifications.

Wavelength	1300 nm
Attenuation rate	0.5 dB/km
Dispersion	3.5 ps/km/nm @ 1310 nm

Cable Performance

Indoor -

Maximum recommended installation load	360-500 pounds
Temperature operating range	-10 to 60 degrees C.

Outdoor -

Maximum recommended installation load	600 pounds
Temperature operating range	-40 to 60 degrees C.
If rodent resistance is required	Use armored cable or install cable in inner duct
Strength members	Use epoxy-fiberglass or flexible fiberglass
Jacket material	Black polyethylene

Chapter 5

Testing

Testing is a critical part of datacom installations. This is different than the usual electrical power installations where testing is very quick and easy. (Hit the outlets with a Wiggy, turn on the lights, maybe use a Megger once in a while.)

Datacom work requires that you not only test each individual conductor, but to document the results and furnish copies of this documentation to the customer. Needless to say, testing datacom cabling takes a lot more time than testing power wiring.

Components

Using Cat 5, 6, or 7 cable alone doesn't guarantee high performance network cabling. Once the cable is properly installed, it is mandatory that the cable be terminated with appropriate connectors and procedures, then using only properly rated (Cat 5, 6, or 7) punchdown blocks, patch panels and patchcords to complete the cable plant. Unless all the appropriate components are used, the system cannot properly be called Category 5, 6, or 7.

What Testing Proves

When someone purchases category 5, 6, or 7 cable and connectors, why should they have to test these parts after installing them? Isn't it enough to just buy quality material and install it by the rules?

We must test, because even if the cable and connectors meet the category 5 specifications, if they are not installed properly, the overall performance that the installed cable run can provide can be substantially below the minimum required by the category 5 definition.

To make matters worse, the effects of poor installation work may not be immediately evident if the first user to be attached to the cable run is only operating at a relatively low speed such as 10 Mbps. Many problems will only become apparent when the user tries to switch up to 100 Mbps or higher - exactly the worst time for such a problem to emerge!

This is not an uncommon problem. Even when high quality cable and connectors are used, a high percentage of all installed cable runs can deliver sub-category 5 performance if the installer is not using the proper techniques. Therefore, it is important to require all category 5 cabling to be field certified after it is installed. It is also important to retest cabling after any changes are made as the network grows or as reconfigurations occur. Even patch cords, typically the shortest part of any cable run, can cause a big problem if not constructed and maintained properly.

Unshielded Twisted Pair (UTP) Testing

Cable/Network	"Level" 1,2	Cat 3	Cat 5
Analog phone	W	W	W
Digital Phone/PBX	1 -NR, 2 -W	W	W
Ethernet	NR	L, W, X, A	L, W, X, A
4 MB Token Ring	NR	L, W, X, A	L, W, X, A
16 MB Token Ring	NR	L, W, X, A	L, W, X, A
100Base-T4	NR	NR	L, W, X, A, PD, S
100Base-Tx	NR	NR	L, W, X, A, PD
100VG AnyLAN	NR	L, W, X, A, S	L, W, X, A, S
TP-PMD/FDDI	NR	NR	L, W, X, A
155 MB/s ATM	NR	NR	L, W, X, A
GB Ethernet	NR	NR	Undefined

Tests:
L = length
W = wire map of connections
X = NEXT (crosstalk)
A = attenuation
S = delay skew
PD = propagation delay
NR = not recommended
Cat 3, 4 & 5 specified by TSB

Table 5-1. Test parameters.

Testing can also find interactions between the cable and connector which are not detectable independently. Some brands of cable and connector can perform poorly when installed together in short (under 60 foot) runs. Field testing is the only way to find these marginal incompatibilities.

How To Start

As a cable installer, you should assure the quality of your materials and work by certifying all installed cable runs at the end of each job using equipment that is specifically designed for testing these data links. Digital multimeters alone are inadequate for this task. While a good multimeter may be able to confirm that signal is getting through from one end of the cable to another, it cannot tell you anything about the quality of the signal. Are there crisp pulses coming through? What is the ratio of attenuation to crosstalk? Unless your tester can do these things well, it is inadequate for data testing.

In addition, it is helpful to have a tester that saves its results, and is set up to transfer them to a computer program for documentation. (This feature will save you a lot of time when completing the project.

Data cabling should be tested each time you install, move or troubleshoot a LAN-attached workstation, so that you can prevent cabling problems from impacting the performance of your high speed network.

Understanding Decibels

In data cabling, most energy and power levels, losses or attenuations, are expressed in the deciBel rather than in the Watt. The reason is simple: Transmission calculations and measurements are almost always made as *comparisons* against a reference: Received power compared to emitted power, energy in versus energy out (energy lost in a connection), etc.

Generally, energy levels (emission, reception, etc.) are expressed in *dBm*. This signifies that the reference level of 0 dBm corresponds to 1 mW of power.

Generally, power losses or gains (attenuation in a cable, loss in a connector, etc.) are expressed in dB.

The unit dB: is used for very low levels.

DeciBel measurement works as follows - a difference of 3 dB equals a doubling or halving of power.

A 3 dB gain in power means that the optical power has been doubled. A 6 dB gain means that the power has been doubled, and doubled again, equaling four times the original power. A 3 dB loss of power means that the power has been cut in half. A 6 dB loss means that the power has been cut in half, then cut in half again, equaling one forth of the original power.

A loss of 3 dB in power is equivalent to a 50% loss. For example: 1 milliwatt of power in, and .5 milliwatt of power out.

A 6 dB loss would equal a 75% loss. (1 milliwatt in, .25 milliwatt out.)

When To Test

The testing of network cables should be done both during the installation process and upon completion of the system. Testing during the installation process helps catch problems while they are still simple to fix. Testing the system upon completion is not only a good practice but is even required by law for communications systems.

Common Cable Test Equipment

The most common testing tools for copper data cabling are the following:

DVM *(Digital Volt Meter).* Measures volts.

DMM *(Digital Multi Meter).* Measures volts, ohms, capacitance, and some measure frequency.

TDR *(Time Domain Reflectometer).* Measures cable lengths, locates impedance mismatches.

Tone Generator and Inductive Amplifier. Used to trace cable pairs, follow cables hidden in walls or ceiling. The tone generator will typically put a 2 kHz audio tone on the cable under test, the inductive amp detects and plays this through a built-in speaker.

Wiremap Tester. Checks a cable for open or short circuits, reversed pairs, crossed pairs and split pairs.

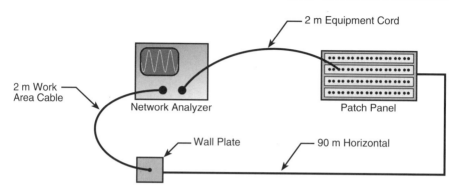

Fig. 5-1. Shown is the test setup for testing network and system performance under high frequencies. This arrangement allows testing in both directions through the network. It also approximates a long-length basic link as defined in TIA/EIA 568A, TSB 67. We conducted separate tests using different makes of patch cables and hardware.

Noise testers, 10Base-T. The standard sets limits for how often noise events can occur, and their size, in several frequency ranges. Various handheld cable testers are able to perform these tests.

Butt sets. A telephone handset that when placed in series with a battery (such as the one in a tone generator), allows voice communication over a copper cable pair. Can be used for temporary phone service in a wiring closet.

Testing UTP Cables

Many of the problems encountered in UTP cable plants are a result of

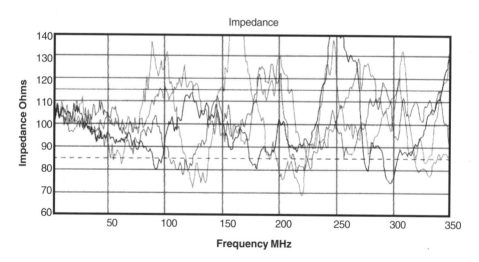

Fig. 5-2. Impedance variation increases at about 50 MHz. (Waves represent the four different pairs within the channel.)

mis-wired patch cables, jacks and cross-connects.

Horizontal and riser distribution cables and patch cables are wired straight through end-to-end; so that pin 1 at one end is connected to pin 1 at the other. (Crossover patch cables are an exception to this rule.) Normally, jacks and cross-connects are designed so that the installer always punches down the cable pairs in a standard order, from left to right: pair 1 (Blue), pair 2 (Orange), pair 3 (Green) and pair 4 (Brown). The white striped lead is usually punched down first, followed by the solid color. The jack's internal wiring connects each pair to the correct pins, according to the assignment scheme for which the jack is designed: EIA-568A, 568B, USOC or whichever.

One common source of problems is an installation in which USOC jacks are mixed with EIA-568A or 568B. When this is done, everything appears to be punched down correctly, but some cables will work and others will not.

Wiremapping

Wiremap tests will check all lines in the cable for all of the following errors:

Open: Lack of continuity between pins at both ends of the cable.

Short: Two or more lines short-circuited together.

Crossed pair: A pair is connected to different pins at each end (example: pair 1 is connected to pins 4 & 5 at one end, and pins 1&2 at the other).

Reversed pair: The two lines in a pair are connected to opposite pins

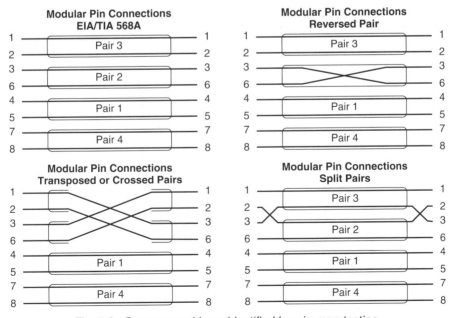

Fig. 5-3. Common problems identified by wiremap testing.

65

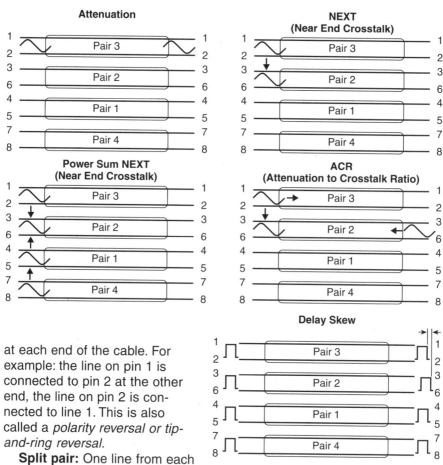

at each end of the cable. For example: the line on pin 1 is connected to pin 2 at the other end, the line on pin 2 is connected to line 1. This is also called a *polarity reversal or tip-and-ring reversal*.

Split pair: One line from each of two pairs is connected as if it were a pair. For example, the Blue and White-Orange lines are connected to pins 4&5, White-Blue and Orange to pins 3&6. The result is excessive Near End Crosstalk (NEXT), which wastes 10Base-T bandwidth and usually prevents 16 Mb/s token-ring from working at all.

Length Tests

Checking cable length is usually done using a time domain reflectometer (TDR), which transmits a pulse down the cable, and measures the elapsed time until it receives a reflection of the signal from the far end of the cable. Each type of cable transmits signals at something less than the speed of light. This factor is called the *nominal velocity of propagation* (NVP), expressed as a decimal fraction of the speed of light. (UTP has an NVP of approximately 0.59-0.65). From the elapsed time and the NVP, the TDR calculates the cable's length. A TDR may be a special-purpose, or may be built into a handheld cable tester.

Testing for Impulse Noise

The 10Base-T standard defines limits for the voltage and number of occurrences per minute of impulse noise occurring in several frequency ranges. Many of the handheld cable testers include the capability to test for this.

Near-End Crosstalk (Next)

To understand NEXT, imagine yourself speaking into a telephone - you can hear the person on the other end and also hear yourself through the handset. Imagine how it would sound if your voice was amplified so it was louder than the other person's. Each time you spoke you could barely hear any sound coming from the other end, due to the contrasting levels of volume. A cable with inadequate immunity to NEXT couples so much of the signal being transmitted back onto the receive pair (or pairs) that incoming signals are unintelligible.

Cable and connecting hardware installed using poor practices can have their NEXT performance reduced significantly.

Attenuation

A signal traveling on a cable becomes weaker the further it travels. Each interconnection also reduces its strength. At some point the signal becomes too weak for the network hardware to interpret reliably. Particularly at higher frequencies (10MHz and up) UTP cable attenuates signals much sooner than does co-axial or shielded twisted pair cable. Knowing the attenuation (and NEXT) of a link allows you to determine whether it will function for a

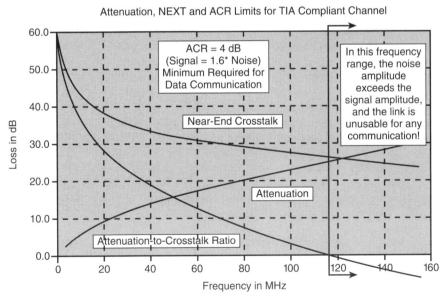

Fig. 5-4. Attenuation-to-crosstalk ratio (ACR).

particular access method, and how much margin is available to accommodate increased losses due to temperature changes, aging, etc.

Testing Optical Fiber

When installing a fiber cable plant, it must be tested, to assure that it will properly perform. The purpose of testing is to make sure that light will pass through the system properly. The main types of optical testing used in the field are the following:

Continuity testing. This is a simple visible light test. Its purpose is to make sure that the fibers in your cables are continuous, that is, that they are not broken. This is done with a modified type of flashlight device and the naked eye, and takes only a few minutes to perform.

Power testing. This is to accurately measure the quality of optical fiber links. A calibrated light source puts infrared light into one end of the fiber, and a calibrated meter measures the light arriving at the other end of the fiber. The loss of light in the fiber is measured in decibels.

OTDR testing. The OTDR is a piece of equipment properly called an *Optical Time Domain Reflectometer*. This device uses light backscattering to analyze fibers. In essence, the OTDR takes a snapshot of the fiber's optical characteristics. It sends a high powered pulse into the fiber and measures the light scattered back toward the instrument. The OTDR can be used to locate fiber breaks, splices and connectors, as well as to measure loss. However, the OTDR method of loss measurement may not give the same value for loss as a source and power meter, due to the different methods of measuring loss. In addition, the OTDR gives a graphic display of the status of the fiber being tested. Another advantage is that it requires access to only one end of the fiber.

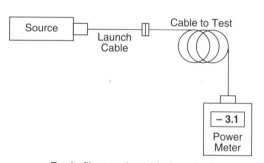

Basic fiber optic cable loss test.

As useful as the OTDR is, however, it is not necessary in the majority of situations. A power meter and source are used to test the loss of fiber optic cable, simulating the way the fibers are used, and measuring the light lost from one end of the cable to the other. In addition, OTDRs

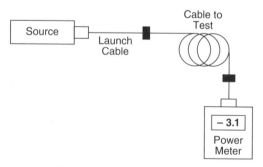

Double-ended cable loss test.
Fig. 5-5.

are quite expensive. Even when they are necessary, many installers prefer to rent them, rather than purchasing them.

Power Meter Testing

Fiber optic power meters measure the average optical power emanating from an optical fiber and are used for measuring power levels and, when used with a compatible source, for loss testing. They typically consist of a solid state detector, signal conditioning circuitry and a digital display of power. To interface to the large variety of fiber optic connectors in use, some form of removable connector adapter is usually provided.

Power meters are calibrated to read in milliwatts, microwatts and/or *dBm* (this references the deciBel to one milliwatt; thus a dBm reading refers to how many dBs the power level is from one milliwatt). Some meters offer a relative dB scale also, useful for laboratory loss measurements.

Although most fiber optic power and loss measurements are made in the range of 0 dBm to -50 dBm, some power meters offer much wider ranges. For testing analog CATV systems or fiber amplifiers, special meters with extended high power ranges are used. Optical power meters have a typical measurement uncertainty of +/-5%.

Visual Cable Tracers and Fault Locators

The most common fault with fiber optic systems is poor connections. Since the light used in systems is not visible, visual inspections will not tell you whether or not the transmitter is operating. This problem can be overcome by injecting the light from a visible source, such as a LED or incandescent light bulb. By doing so, you can visually trace the fiber from transmitter to receiver, and verify that the light is traveling the correct course. The test instruments that are used to inject visible light are called *visual fault locators*.

If a powerful enough visible light, such as a HeNe (Helium-Neon) or visible diode laser is injected into the fiber, high loss points can be made visible. This method will work on buffered fiber and even jacketed single fiber cable if the jacket is not opaque to the visible light. The yellow jacket of singlemode fiber and orange of multimode fiber will usually pass the visible light. Most other cable colors, especially black and gray, will not work with this technique, nor will most multi fiber cables. However, many cable breaks, macro bending losses caused by kinks in the fiber, bad splices etc. can be detected visually.

Since the loss in the fiber is quite high at visible wavelengths, on the order of 9-15 dB/km, this instrument has a short range, typically 3-5 km, which is more than enough for virtually all indoor installations.

Microscopes

Generally the first test performed once the fibers are pulled into place is a visual inspection with a microscope. This is done when preparing fiber ends for termination.

The microscopes used for testing fibers have stages that are modified

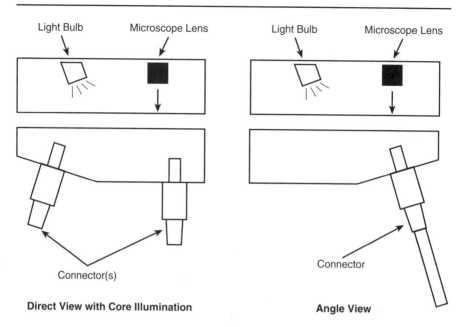

Fig. 5-6. Inspecting fiber connection with microscope.

to hold the fiber or connector in the field of view. Fiber optic inspection microscopes vary in magnification from 30 to 800 power, with 30-100 power being the most widely used range. Cleaved fibers are usually viewed from the side, to see *breakover* and *lip*. Connectors are viewed end-on or at a small angle to find polishing defects such as scratches.

A well made connector will have a smooth, polished, scratch free finish, and the fiber will not show any signs of cracks or pistoning (where the fiber is either protruding from the end of the ferrule or pulling back into it).

The proper magnification for viewing connectors is generally accepted to be 30-100 power. Lower magnification, typical with a jeweler's loupe or pocket magnifier, will not provide adequate resolution for judging the finish on the connector. Too high a magnification tend to make small, ignorable faults look worse than they really are. A better solution is to use medium magnification, but inspect the connector three ways:

1. By viewing directly at the end of the polished surface with side lighting.
2. By viewing directly with side lighting and light transmitted through the core.
3. By viewing at an angle with lighting from the opposite angle.

Viewing directly with side lighting allows determining if the ferrule hole is of the proper size, the fiber is centered in the hole and a proper amount of adhesive has been applied. Only the largest scratches will be visible this way, however. Adding light transmitted through the core will reveal cracks in the end of the fiber that are caused by pressure or heat during the polish process.

Viewing the end of the connector at an angle, while lighting it from the opposite side at approximately the same angle, will allow the best inspection for the quality of polish and possible scratches. The shadowing effect of angular viewing enhances the contrast of scratches against the mirror smooth polished surface of the glass.

One needs to be careful in inspecting connectors, however. The tendency is to be overly critical, especially at high magnification. Only defects over the fiber core are a problem. Chipping of the glass around the outside of the cladding is not unusual and will have no effect on the ability of the connector to couple light in the core. Likewise, scratches only on the cladding will not cause any loss problems.

Continuity Testing

Continuity testing is the most fundamental fiber optic test. It is usually performed when the cable is delivered to the job site (and certainly before installation) to insure that no damage has been done to the cable during shipment to the work site.

This testing is commonly done with a visible light source, which can be an incandescent light bulb, HeNe laser at 633 nm or a LED or diode laser at 650 nm, a wavelength (that is, color) which can be seen by the eye. HeNe laser instruments are usually tuned to an output power level of just less than 1 mW, making then Class II lasers which do not have enough power to harm the eye, but do have enough power to be easily seen over about 4 km of fiber and even find fiber microbends or breaks by viewing the light shining from the break through the yellow or orange jacket used on most single fiber cables. In most cases, a high quality flashlight, shined directly into a good fiber end, work quite well.

Optical Time Domain Reflectometer

The optical time domain reflectometer (OTDR) is one of the most powerful types of optical fiber testers. It uses the phenomena of back scattered light to analyze fibers, find faults and optimize splices.

If one assumes that the average amount of backscatter (called the *backscatter coefficient*) is constant, the OTDR can be used to measure loss as well as locate fiber breaks, splices and connectors. In addition, the OTDR gives a graphic display of the status of the fiber being tested. And it offers another major advantage over the source/power meter, in that it requires access to only one end of the fiber.

The uncertainty of the OTDR measurement is heavily dependent on the backscatter coefficient. Tests have shown that OTDR splice loss measurements may have an uncertainty of up to 0.8 dB. OTDRs must also be matched to the fibers being tested in both wavelength and fiber core diameter to provide accurate measurements. Thus many OTDRs have modular sources to allow substituting a proper source for the application.

While most OTDR applications involve finding faults in installed cables or checking splices, they are also very useful in inspecting fibers for manufacturing faults. They are very nice for taking a "snapshot" of a fiber run.

The Limitations of OTDR Use

With the OTDR, one can measure loss and distance. To use them effectively, it is necessary to understand their measurement limitations. The OTDR's distance resolution is limited by the transmitted pulse width. As the OTDR sends out its pulse, crosstalk in the coupler inside the instrument and reflections from the first connector will saturate the receiver. The receiver will take some time to recover, causing a nonlinearity in the baseline of the display. It may take the time equivalent to 100-1000 meters before the receiver recovers. It is common to use a long fiber cable called a pulse suppressor between the OTDR and the cables being tested to allow the receiver to recover completely.

The OTDR also is limited by the pulse width in its ability to resolve two closely spaced features. Long distance OTDRs may have a minimum resolution of 250 to 500 meters, while short range OTDRs can resolve features 5-10 meters apart. This limitation makes it difficult to find problems inside a building, where distances are short. A *visual fault locator* is generally used to assist the OTDR in this situation.

When measuring distance, the OTDR has two major sources of error not associated with the instrument itself: the velocity of the light pulse in the fiber and the amount of fiber in the cable. The velocity of the pulse down the fiber is a function of the average index of refraction of the glass. While this is fairly constant for most fiber types, it can vary by a few percent. When making cable, it is necessary to have some excess fiber in the cable, to allow the cable to stretch when pulled without stressing the fiber. This excess fiber is usually 1-2%. Since the OTDR measures the length of the fiber, not the cable, it is necessary to subtract 1-2% from the measured length to get the likely cable length. This is very important if one is using the OTDR to find a fault in an installed cable, to keep from looking too far away from the OTDR to find the problem. This variable adds up to 10-20 meters per kilometer, therefore it is not ignorable.

When making loss measurements, two major questions arise with OTDR measurement anomalies:
1. Why OTDR measurements differ from an optical loss test set, which tests the fiber in the same configuration in which it is used.
2. Why measurements from OTDRs vary so much when measured in opposite directions on the same splice. (Sometimes, one direction shows a "gain", not a loss.)

To understand the problem, it is necessary to consider again how OTDRs work. They send a powerful laser pulse down the fiber, which suffers losses as it proceeds. At every point on the fiber, part of the light is scattered back up the fiber. The back scattered light is then attenuated by the fiber again, until it returns to the OTDR and is measured.

Note that three factors affect the measured signal: attenuation outbound, scattering and attenuation inbound.

It is commonly assumed that the backscatter coefficient is a constant, and therefore the OTDR can be calibrated to read attenuation. The backscatter coefficient is, in fact, a function of the core diameter of the fiber, and the material composition of the fiber. Thus a fiber with either higher attenuation due to scattering or larger core size will produce a larger backscatter signal.

Accurate OTDR attenuation measurements depend on having a constant backscatter coefficient. Unfortunately, this is often not the case. Fibers which have tapers in core size are common, or variations in diameter are a result of variations in pulling speed as the fiber is being made. A small change in diameter (1%) causes a larger change in cross sectional area which directly affects the scattering coefficient and can cause a large change in attenuation (on the order of 0.1 dB). Thus fiber attenuation measured by OTDRs may not be uniform along the fiber, and can produce significantly different losses in opposite directions.

The first indication of OTDR problems for most users occurs when looking at a splice and a "gain" is seen at the splice. Common sense tells us that passive fibers and splices cannot create light, so another phenomenon must be at work. In fact, a "gainer" is an indication of the difference of backscatter coefficients in the two fibers being spliced.

If an OTDR is used to measure the loss of a splice and the two fibers are identical, the loss will be correct, since the scattering coefficient is the same for both fibers. This is exactly what you see when breaking and splicing the same fiber, the normal way OTDRs are demonstrated.

If the receiving fiber has a lower backscatter coefficient than the fiber before the splice, the amount of light sent back to the OTDR will decrease after the splice, causing the OTDR to indicate a larger splice loss than actual.

If one looks at this splice in the opposite direction, the effect will be reversed. The amount of back scattered light will be larger after the splice and the loss shown on the OTDR will be less than the actual splice loss. If this increase is larger than the loss in the splice, the OTDR will show a gain at the splice, an obvious error. As many as one-third of all splices will show a gain in one direction.

The usual recommendation is to test with the OTDR in both directions and average the reading, which has been shown to give measurements accurate to about 0.01 dB. However, this negates the most useful feature of the OTDR, the ability to work from only one end of the fiber.

When To Use The OTDR

Once upon a time, OTDRs were used for all testing of installed cable

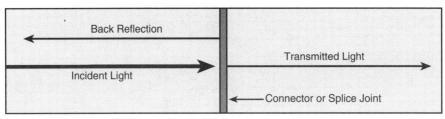

Fig. 5-7. Back reflection.

plants. In fact, printouts or pictures of the OTDR traces were kept on record for every fiber in every cable. The power meter and source have replaced the OTDR for most final qualification testing since the direct loss test gives a more reliable test of the end to end loss than does an OTDR.

However, the OTDR may need to be used to find bad splices or optical return loss problems in connectors and splices in a singlemode cable plant. Only with an OTDR can return loss problems be located for correction. Typical back reflection test sets only give a total amount of backscatter or return loss, not the effects of individual components, which is necessary to locate and fix the problem.

The OTDR can also be used to find bad connectors or splices in a high loss cable plant, if the OTDR has high enough resolution to see short, individual cable assemblies. However, if the cables are too short or the splices too near the end of the fiber (as is often the case in pigtails spliced onto singlemode fiber cables), the only way to localize the problem is to use a visual fault locator, preferably a high-power HeNe laser type, which can shine through the jacket of typical yellow or orange PVC-jacketed single fiber cables. This method of fault location is easiest if single-fiber cables use yellow or orange jackets that are more translucent to the laser light.

Chapter 6

Outside Plant Installations

Outside plant (OSP) installations are installations of cabling outdoors. (As opposed to *inside* plant (ISP) cabling, which refers to cables installed indoors.) Outside plant cabling refers to both cables buried in the ground, and cables installed aerially on poles, the sides of buildings, etc.

Outside plant work and inside plant work have historically been done by different groups of people. This tradition goes back to the operating methods of telephone companies, where there were specific crews that installed inside wiring, and other crews that installed outside wiring. Sometimes these crews were even in different union locals. Most of the time, they were different classes of employees. But now that non-AT&T contractors are installing most data communication circuits, there are a lot more people who work both indoors and outdoors.

Another relic of the telephone company that you may encounter is that you may hear outside plant cables referred to as *black cables*. In the days of the AT&T monopoly, all outdoor cables were colored black - thus the name. (Even today, most outdoor cables are still colored black.)

Burying Cable

There are a number of reasons why people prefer to bury cables, rather than to place poles and run the cable aerially (which is frequently a less-expensive option). One reason is that a direct burial cable is virtually free from storm damage and has lower maintenance costs than aerial cable. In addition, aerial installations often lack aesthetic appeal, and in some communities, are even prohibited.

Although any cable can be buried in the earth, a cable that is specifically designed for direct burial will have a longer life. A cable with a high-density polyethylene jacket is particularly well equipped for direct burial because it can stand up well to compressive forces. High-density polyethylene is both non-porous and non-contaminating, and provides complete protection against normal moisture and alkaline conditions. In many direct-buried cables, an additional moisture barrier of water-blocking gel is applied under the jacket. If water should penetrate the jacket, it would not be able to travel under the jacket, damage would be localized and more readily repaired. Because buried cable is thermally insulated by the earth, its year-round temperature will only vary a few degrees. If a buried cable is not damaged the attenuation will be constant for the useful life of the cable.

The tools required for burying cable are well-known, consisting mainly of trenchers, backhoes, and shovels.

Underground Cable Installation

When installing underground cable, these are the things that you should take into account:

• Because the outer jacket is the cable's first line of defense, any steps that can be taken to prevent damage to it will go a long way toward maintaining the internal characteristics of the cable.

• It is generally best to bury the cable in sand or finely pulverized dirt, without sharp stones, cinders or rubble. If the soil in the trench does not meet these requirements, tamp four to six inches of sand into the trench, lay the cable, and tamp another six to eleven inches of sand above it. A creosoted or pressure-treated board placed in the trench above the sand, prior to back filling, will provide some protection against subsequent damage that could be caused by digging or driving stakes.

• Lay the cable in the trench with some slack. A tightly stretched cable is likely to be damaged as the fill material is tamped.

• Examine the cable as it is being installed to be sure the jacket has not been damaged during storage, by being dragged over sharp edges on the pay-off equipment, or by other means.

• In particularly difficult installations, such as in rubble or coral, or where paving is to be installed over the cable, a polyethylene water pipe, which is available in long lengths and several diameters, may be buried and used as a conduit. This pipe protects the cable and usually makes it possible to replace cable that has failed without digging up the area.

• It is important that burial is below the frost line to avoid damage by the expansion and contraction of the earth during freezing and thawing.

• The National Electrical Code (NEC) states specific requirements for cables to be buried underground. The NEC specifies 24 inches as the minimum burial depth for 0-600 volt nominal applications. If an installation must meet all NEC requirements, a local inspector should be consulted during the planning phase.

• Study the application, determine if any failure of the cable will result in a hazardous condition, and choose your cable accordingly.

Underground Raceways

While installing directly-buried cables are certainly common, there are many times when underground cables are installed in raceways. The advantages in this are superior protection, and the ability to add conductors at a future date with a minimum of expense. Obviously, it is far more expensive to install cables in conduit than it is to install directly-buried cables.

When installing underground conduits, it is important to install more raceway than you need at the time of installation. This allows you to install more cables in the future. Pulling more cables in a conduit that is already partially-filled is not generally a viable plan - working in pre-filled conduits is frequently more difficult than planned.

Another thing to do when installing underground conduits is to install multiple innerducts inside one large conduit. This allows you to have three or four separate mini-raceways inside of the one large raceway.

This method often eliminates the need for pull rope and reduces the amount of lubricant required.

Blown-In Fiber

When working with multiple innerducts inside a conduit, one of the quickest methods of installing optical fiber is to blow it in. When done properly, this method works very well.

The process work as follows:
- A compressor with about 140 PSI supplies air to the jetting machine. Air is supplied to one innerduct to clean out any possible debris.
- Small foam sponges are shot through each innerduct to further clean the system.
- A few ounces of lubricant are added to the already pre-lubricated PVC innerducts.
- A tarp is laid out to keep the cable clean before innerduct entry.
- A small hollow "bullet" is attached to the end of the cable for alignment.
- The fiber cable is blown into the duct at up to 200 feet per minute at a pressure of 110 PSI to 120 PSI. (Speeds of 300+ feet are obtainable using pneumatic units.)

Routing of Outdoor Aerial Circuits

The code's rules or the routing of aerial coaxial and communications

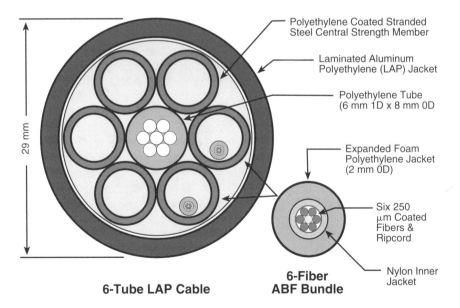

Fig. 6-1. Typical tube cable cross section contains six inner tubes, through which you can blow a variety of fiber bundles.

circuits are that such cables must:
- Be run below power conductors on poles.
- Remain separated from power conductors at the attachment point to a building.
- Have a vertical clearance of 8 feet above roofs (there are several exceptions).
- Coax cables run on the outside of buildings must be kept at least 4 inches from power cables (but not conduits).
- They must be installed so that they do not interfere with other communications circuits; generally, this means that they must be kept far enough away.
- Coaxial and communication cables must be kept at least 6 feet from all lightning protection conductors, except if such spacing is very impractical.

Messenger Cables

The vast majority of outdoor runs of cables are run with messenger cables. (Or specially designed to be messenger cable assemblies themselves.) In article 820 of the NEC, the only requirements made for such runs is that the runs be attached to a messenger cable that is acceptable for the purpose, and has enough strength for the load to which it will be subjected (such as the weight of ice or snow, or wind tensions). Messenger cables are covered in more detail in article 321 of the NEC.

Underground Cable Locators

One tool that you will have to use if you do much underground work is a cable locator.

Cable locators generally consist of two parts: a transmitter and a receiver. The transmitter puts an electrical signal onto the cable or pipe being traced, while the receiver picks up that signal, allowing the locator operator to trace the signal's path and follow the cable being located.

The electromagnetic field created by the transmitter can usually be set to a specific frequency, depending on the type of conductor in the cable. Frequency choices can range from less than 1 kilohertz to about 480 kHz. With this range of frequencies, it is important to start out at the lowest frequency, and if that frequency works, stick with it. Lower frequencies seem to bleed over less (lower crosstalk), and stay in the conductor to which the transmitter is connected.

While the basic technology for locating hasn't changed much, the ways for getting the signal onto the cable have. In addition to passive signals, which do not require the use of a transmitter, the three most common methods of sending signals are direct connect, general induction, and inductive coupling. In the inductive coupling method, the cable must be grounded to form a complete circuit path.

As its name implies, the direct-connect method allows you to physically attach your transmitter to the cable to be located. That may mean connecting at a cabinet or a pedestal and then gaining access to the shield that surrounds the cable, which is usually grounded at this point.

If directly attaching to the cable is impossible, then the induction method may be the best choice. Here, you place the transmitter on the ground directly over the cable. Once the transmitter is turned on, it induces a signal into any nearby conductor within its range. This, of course, can lead to problems if there are multiple cables buried within close proximity, since the signal could be picked up on a cable other than the one you are trying to trace.

Although inductive coupling doesn't let the user directly connect to the cable, it provides a higher level of confidence than does the general induction method. It uses a donut-shaped coupling device that surrounds the cable and emits a signal onto the cable.

Directly Buried Optical Cables

Only armored or other heavy-duty cables are directly buried. Directly buried cables are frequently exposed to ground water, and should be water-blocked in the core and within the primary sheath to avoid water migration along the cable. A number of methods may be used.

Optical cables should be buried below the frost line for the geographic area in which they are installed; exceptions may be made for locations where permafrost exists. Steel conduit, metal braids, or substantial armor are used where cables may be subject to rodents or gophers.

When cables are plowed into place, it is important that you use only cables designed for that purpose.

Where freezing water presents a threat, cables ought to be completely filled to prevent the entrance or migration of water.

Optical Aerial Cables

When optical cables are installed aerially, they are supported by a messenger wire as required by the National Electrical Safety Code and Article 321 of the National Electrical Code.

With few exceptions, aerial cables are round, loose tube, outside plant cables, and must have temperature ranges (specified by the manufacturer) that correspond to the environment in which they will be installed. Cables for long outdoor aerial runs must be temperature stabilized.

To account for thermal expansion and contraction, all outdoor aerially-installed cables must follow the requirements of the National Electrical Safety Code.

Hybrid Networks

Hybrid networks are those built with both copper cabling (category 5 or higher) and optical fiber. There are two common types of hybrid networks, backbone systems, and campus systems.

Backbone Systems

In almost backbone systems, copper cabling is used for the horizontal cable runs (runs contained on a single floor of a building), and optical fiber cables are used for the vertical runs (cable run from the bottom to the top of the building, between floors).

In this arrangement, the vertical cabling (fiber) carries all of the data traffic that travels from floor to floor. Fiber is used here because of its much higher capacity than copper cabling.

Campus Systems

Campus systems are composed of several buildings (typically with fiber backbones in them) connected together with optical fibers run underground between the buildings.

Types of Fiber

In most cases, multimode optical fibers are used for backbones or campus links. Where multimode fiber provides enough bandwidth (signal transmission capabilities), it is commonly used. Multimode fiber systems are generally the least expensive kind, since they can work well with inexpensive (LED) receivers and detectors.

For longer underground runs (usually over a few kilometers), singlemode fiber is used. Singlemode is a bit more expensive to use than multimode fiber, since it must use the more expensive LASER light sources. However, singlemode fiber has an almost unlimited bandwidth. (Current long-distance telephone systems operate at up to 100 gigabits per second, and laboratory systems operate at many times that speed.)

Installation

In this section, we will not cover the design, installation, or testing of horizontal (copper) cable runs, since we have already covered them in great detail in other places. Our concern here is the fiber backbone or campus link.

Installation of vertical backbone data cabling between floors of a building is typically composed of a riser-rated tight-buffered fiber optic multimode cable, pulled between the communication closets on each floor. Installation of backbone data cabling between buildings will include loose-tube, multimode, optical cable, pulled between the communication closets in each building. Fiber is terminated on each floor or in each building and installed in interface units in each closet. Duplex fiber jumper cables are used for connection between the interface units and hubs on each floor.

Types of Fiber Cables

Vertical data cabling is usually an indoor riser-rated plenum-rated multimode fiber optic graded index glass with a 62.5 micrometer core, 125 micrometer cladding, and a numerical aperture of .025 to 0.30. The bandwidth must be 160MHz or greater at 850 nm and have an attenuation of 3.75 dB per kilometer or less at 850 nm. Almost always, FDDI grade fiber is used. The fiber must be a continuous strand certified on the reel.

Campus links between buildings the backbone data cabling will be outdoor armored aerial or direct burial loose-buffered fiber optic multi-mode fiber optic graded index glass with a 62.5 micrometer core, 125 micrometer cladding, and a numerical aperture of .025 to 0.30. These cables, also called *trunk cables*, must be suitable for underground conduit or aerial installations, and rated for vertical installations up to 200 feet with no degradation in performance. The armored cables will require grounding and bonding.

Jumpers are generally 3-meter duplex multimode ST-ST 62.5/125 fiber optic jumpers. These jumpers are all-dielectric and have built-in strain relief with an aramid yarn strength member and overall PVC jacket for each fiber.

Industry Cable Standards

Aside from NEC fire-rating classifications such as general-duty, riser-rated cable, and plenum-rated cable, the physical construction of optical cables is not governed by any agency. It is up to the designer of the system to make sure that the cable selected will meet the application requirements. Five basic cable types have however emerged as de facto standards for a variety of applications:

Simplex and zip cord: One or two fibers, tight-buffered, Kevlar reinforced and jacketed. Used mostly for patch cord and backplane applications.

Distribution cables: (Also known as tightpack cables.) Up to several tight-buffered fibers bundled under the same jacket with Kevlar reinforcement. Used for short, dry conduit runs, riser and plenum applications. These cables are small in size, but because their fibers are not individually reinforced, these cables need to be terminated inside a patch panel or junction box.

Breakout cables: These cables are made of several simplex units, cabled together. This is a strong, rugged design, and is larger and more expensive than the distribution cables. It is suitable for conduit runs, riser and plenum applications. Because each fiber is individually reinforced, this design allows for a strong termination to connectors and can be brought directly to a computer backplane.

Loose tube cables: These are composed of several fibers cabled together, providing a small, high fiber count cable. This type of cable is ideal for outside plant trunking applications. Depending on the actual construction, it can be used in conduits, strung overhead or buried directly into the ground.

Hybrid or composite cables: There is a lot of confusion over these terms, especially since the 1993 United States National Electrical Code switched their terminology from "hybrid" to "composite".

Under the new terminology, a *composite* cable is one that contains a number of copper conductors properly jacketed and sheathed depending on the application, in the same cable assembly as the optical fibers.

This situation is made all the more confusing since there is another type of cable that was formerly called composite. This type of cable contains only optical fibers, but have two different types of fibers: Multi-mode and single-mode.

Remember that there is confusion over these terms; with some people using them interchangeably. At this point the proper terminology is the following:

A *composite* cable is a fiber/copper cable.

A *hybrid* cable is a fiber/fiber cable.

Connectors and Splices

Connecting and splicing optical fibers is generally the most labor-intensive part of the installation process. Pulling the cables into place is relatively easy (requiring just a bit more care than power cables), and other parts of the installation are comparably simple. But connecting these glass fibers correctly requires time, special tools, and specific skills.

All fiber joints must meet two criteria. They must be:

1. *Mechanically strong:* Fiber connections must be capable of withstanding moderate to severe pulling and bending tests.
2. *Optically sound with low loss:* Since the purpose of fiber is to transmit light, the fiber joint must transmit as much light power as possible with as little loss and back reflection as can be designed into the joint.

Fiber connections fall generally into two categories: the permanent or fixed joint which uses a fiber *splice*, and the non-fixed joint which uses a fiber optic *connector*.

Splices are used as a permanent connection. Typical uses include reel ends, for pigtail vault splices and at distribution breakouts. The criteria for good fiber splices are low loss and high mechanical strength. Additional considerations are expense per splice and possible reusability of the splice itself.

Fiber optic connectors are used as a termination for inside cables, outside cables as they terminate in a central office, for interfaces between terminals on LANs, for patch panels, and for terminations into transmitters and receivers.

Whether one joins fibers using splices or connectors, one negative aspect is always common to both methods - signal loss. This loss of light power at fiber connections is called *attenuation*, measured in decibels. The decibel is a mathematical logarithmic unit describing the ratio of output power to input power.

Breakout Kits

Before the installation of connectors onto a fiber cable, a *breakout kit* may have to be installed. These breakout kits are a simple box (into which the cable enters), fitted with semi-rigid tubes, through which the individual fibers exit. The tubes provide mechanical protection for the unprotected fibers, which will usually run from the breakout point to individual connectors or splices.

This procedure is not necessary on "breakout cables" having 2 mm buffered fibers, but will be required on 250, 500 or 900 micron tight buffer cables. The break out kit consists of 2 mm buffer tubing over 900 micron inner tubing. The bare fibers are inserted into these buffer tubes to provide handling protection and strength when mounted onto connectors.

Manholes, Vaults and Poles

A common hazard when optical fiber is installed underground involves placing cables in an unfamiliar environment. Since runs of fiber frequently share facilities with power conductors (whether on poles or in underground raceways), installers will often need to work in these locations. When doing so, it is essential that you know the proper safety requirements for these areas.

Fiber may be safe by itself, but if you are in a confined manhole, or on a pole, you have to be prepared for every hazard that exists there, including high voltage.

It is also critical to install service loops of spare cable in these types of installations. Typically, a 30 foot loop of spare cable is recommended.

Fiber Optic Standards

For any technology to be widely used, a set of common standards must first be in place, which will be adhered to by all the players in the market place. For example, if audio components such as tuners, amps, CD players, and tape decks could not simply plug together, they would be far more difficult to use, and far less popular. The standardization of the audio market means that all audio manufacturers use the same types of jacks and plugs, and send their signals at compatible voltages.

Fiber optic standards today are pretty well in place, and do not pose any great difficulty to the installer. However, once you begin to design or troubleshoot systems, you will have to pay attention to the standards upon which your system is based. When comparing fibers, you must verify which tests are used on each fiber. In other words, the only real problem you will have is making sure that you are comparing "apples and apples".

You will find that some manufacturers of optical fiber products will print testing standards in their catalogs (especially some cable manufacturers), and others will not. There will likely be times when you will have to make a few phone calls to verify some information.

Our existing optical fiber standards include component standards (how components must be built), network standards (how networks must be put together), test methods and calibration standards (how meters must be set). Primary standards for measuring optical power, attenuation, bandwidth and the physical characteristics of fiber are also required.

These standards are developed by an alphabet soup of groups, including the ANSI, IEEE, IEC, EIA, IEC, and others. Primary standards are developed by national standards laboratories such as NIST (National Institutes of Standards and Technology, formerly the US National Bureau of Standards).

In addition to formal standards, there is a second set of rules that have

been adopted by the players n the optical fiber market place. These *de facto* standards include the following:
- Standard wavelengths for optical fiber systems are 850nm (nanometers), 1300nm, and 1550 nm. (For those who are unfamiliar, the various wavelengths are simply slightly different colors of light. Rather than calling them "deep red, slightly deep red, etc., we state the wavelength of each color, which we measure in nanometers.)
- Most fiber optic telecom systems in the US are based on single mode fiber and 1300 nm laser light with a very pure color. They are, however, rather expensive.)
- Most fiber data networks are multimode systems, powered by 850 nm LED light sources. (Light Emitting Diodes are not as powerful as lasers, and can not put out a pure color. They are nonetheless widely used, being very affordable.)
- Type ST connectors are usually used, but are being slowly replaced by SC connectors.
- Although there are several types of multimode fibers that have been used - 50/125, 62.5/125, 85/125 and 100/140 (core/clad measurements, in microns), 62.5/125 fiber is now almost exclusively used.

Choosing Fibers

In the design of optical fiber data transmission systems, the obvious first step is to determine the type of fiber necessary. There are two primary factors that must be considered in choosing fiber:

Bandwidth. Bandwidth is the measure of the data-carrying capacity of the fiber. The greater the bandwidth, the more information that can be carried. Bandwidth is expressed in a frequency-distance figure, usually as megahertz per kilometer (MHZ/km). For example, a 100 MHZ/km fiber can move 100 MHZ of data up to one kilometer or 200 MHZ of data up to one half of a kilometer.

Transmission loss. Besides changes in the light pulse due to frequency and bandwidth limitations, the level of light is reduced as it travels through the fiber. This loss is due to the fiber content itself and bends in the fiber, losses at connections, losses at splices, and losses at couplings. The losses are expressed in decibels per kilometer (dB/km). dB/km is the attenuation coefficient of the fiber itself.

The bandwidth requirements for a data communication system will be determined by the equipment that will be served. The distances required for transmission will also be determined by the requirements and placement of the equipment being served.

To determine the capacity of a given type of fiber for a proposed system, it is necessary to determine both the speed of the data transmission, measured in megahertz, and the distance of transmission. Then, reference is made to a chart such as the one which follows. The far right-hand column shows megahertz per kilometer transmission characteristics for each type of fiber.

For example, if you must send 150 megahertz over a distance of 2 kilometers, a fiber with a rating of at least 300 megahertz per kilometer is required. Thus, a Step Index glass fiber, (having a bandwidth rating of 20 megahertz per kilometer) would be inadequate, but a multi-mode fiber having a rating of 600 MHZ/km would do very well.

Graded index fiber is often used where large core size and efficient coupling of source power is more important than low loss and high bandwidth. It is commonly used in short, low speed data links. It may also be used in applications where radiation is a concern, since it can be made with a pure silica core that is not readily affected by radiation.

While there have been four graded index multimode fibers used over the history of fiber optic communications, one fiber now is by far the most widely used, 62.5/125. Virtually all multimode *datacom* (data communications) networks use this fiber. The first multimode fiber widely used was 50/125, first by the telephone companies who needed its greater bandwidth for long distance phone lines. Its small core and low *NA* (numerical aperture) made it difficult to couple to LED sources, so many data links used 100/140 fiber. It worked well with these data links, but its large core made it costly to manufacture, and its unique cladding diameter required connector manufacturers to make connectors specifically for it. These factors led to its declining use. The final multimode fiber, 85/125 was designed by Corning to provide efficient coupling to LED sources but use the same connectors as other fibers. However, once IBM standardized on 62.5/125 fiber for its fiber optic products, the usage of all other fibers declined sharply.

The telephone companies switched to single-mode fiber a number of years ago because of its better performance at higher bit rates and its lower loss. This allows faster and longer unrepeated links for long distance telecommunications. Virtually all telecom applications use single-mode fiber. It is also used in CATV work (*CATV* is short for Community Antenna Television, which was the original version of cable television; in modern usage CATV refers to cable television), since analog CATV networks use laser sources designed for single-mode fiber. Other high speed networks are using single-mode fiber, either to support gigabit data rates or long distance links.

Typical Bandwidth-Distance Products

Type of fiber	Wavelength	Bandwidth-distance product
Step-index plastic	660 nm	5 MHZ-km
multimode step index	850 nm	20 MHZ-km
multimode graded index	850 nm	600 MHZ-km
multimode graded index	1300 nm	1000-2500 MHZ-km
singlemode	1300 nm	over 300,000 MHZ-km

Choice of Cables

Once the proper type of fiber is chosen, the next step is to choose the type of cable to enclose and protect the fiber. Since fiber is made of glass (except plastic fiber, which is still infrequently used), it is fragile, and the cable with which it is used is very important. Fortunately, fiber optic cable technology has advanced rapidly, resulting in modern cables that are very durable; in many cases they are considerably stronger than copper cables.

Aside from the basic considerations, the following factors are essential in choosing a proper optical cable:

Cable jacketing. The materials used for the outer jacket of fiber optic cables not only affect the mechanical and attenuation properties of the fiber, but also determine the suitability of the cable to different environments, and its compliance to various NEC and UL requirements.

A cable which will be exposed to chemicals, can utilize an inert fluorocarbon jacket such as Kynar, PFA, Teflon FEP, Tefzel, or Halar. These materials are suitable for a very wide range of applications, although they may be too stiff for some industrial applications.

Aerospace applications require that the cables be able to withstand a wide temperature range, and be routed through the cramped environment of an aircraft. These cables are frequently rated for continuous operation from -65 degrees C to +200 degrees C, are less than 1/10 inch in size, and can sustain a bend radius of ½ inch.

Fire safely is a major issue. Cables used in an industrial environment, such as a power plant, are usually placed in horizontal trays. Several cable trays may be stacked in close proximity. In the event of a fire, both horizontal fire propagation and the ignition of lower cable trays by the dripping of flaming outer jacket material must be prevented. An irradiated Hypalon or XLPE jacket will meet the flame spread requirements (IEEE-383, 1974). When exposed to a flame, the cable will char rather than melt and drop burning material, thus preventing the ignition of cables in lower trays.

Inside-premises cables must meet the requirements of the NEC Article 770. The outer jacket selection is essential to ensure compliance to the flame and smoke requirements. (The National Electrical Code's Article 770 governs optical fiber installations.)

Environmental & mechanical factors. Aside from buffer type, the jacketing system and the flammability requirements, the cable design must also be based on the mechanical and environmental conditions that will be encountered throughout the system's life span.

A cable that will be pulled through conduits, ducts or cable trays will have to incorporate a number of strength members and stiffening elements to add tensile strength and to prevent sharp bends from damaging the fibers. The addition of Kevlar increases the cable tensile strength. It can either be braided or longitudinally applied underneath the cable or fiber component jackets. The central strength member also serves both as a filler around which the fiber components are cabled, and as a strength member when it incorporates steel, Kevlar or epoxy glass rods. Another function of the epoxy glass central member is to act as an anti-buckling component, counteracting the shrinkage of the jacketing elements at low temperatures, preventing microbends in the fibers. An epoxy glass rod central member should always be used in cables that may be exposed to temperatures below 0 degrees C.

General Factors

The general factors to be considered when choosing an optic fiber cable plant are as follows:
1. Current and future bandwidth requirements.
2. Acceptable attenuation rate.
3. Length of cable.
4. Cost of installation.
5. Mechanical requirements (ruggedness, flexibility, flame retardance, low smoke, cut- through resistance).
6. Code and safety requirements.
7. Signal source (coupling efficiency, power output, receiver sensitivity).
8. Connectors and terminations.
9. Cable dimension requirements.
10. Physical environment (temperature, moisture, location).
11. Compatibility with any existing systems.

The design process involves sitting down with a diagram of the system being designed, any and all specifications, and pen and paper. Then, each of these factors is considered, compared, and analyzed, to formulate a final design.

Chapter 7

Telephone Systems and Data Transmission

Data transmission over the common telephone system was almost unthought of only ten years ago. Now, it is a huge business, with tens of millions of people doing this every day. In fact, data transmission is becoming a big part of the telephone business, and will probably be more important than voice transmission in a few years. So, there is a world of opportunity open to those who understand this work, and can apply their knowledge intelligently.

But telephone business is also full of change and controversy. Things are changing. Historically, telephone companies have been government-protected monopolies. In other words, you cannot just open up your own phone company; the federal government decides who can have a phone company, and which areas they can serve. In the early 1980s, the government allowed other companies to compete with AT&T in long-distance service. That opened the way for MCI, Sprint, and much lower long-distance rates.

But local-area service is still a protected monopoly. Furthermore, the phone companies see data communication as a serious threat, and are trying to delay it as best they can. And the phone companies are correct - data is a threat to them (we'll explain why in a moment). So, they allow their customers very little access to high-speed lines and make them pay very high prices for it. In addition, they have teams of lawyers working on maintaining their monopolies. On the other hand, the computer industry is going nuts, trying to get the phone companies to loosen-up, and allow people to get high-bandwidth lines. Without affordable high-speed lines, the best features of computers cannot reach the home. The computer guys are starting to get serious about breaking up the monopolies, and some of them can throw enough money around Washington to influence votes.

The end of the phone company?

The future of the phone company is not especially bright. The reason is that their networks are not designed for the world of 1930, not for the 21st century. Data traffic follows entirely different rules.

Telephone companies have billions of dollars invested in switched networks. This allows them to charge you every time your telephone line is switched and connected to another line. The switching system is what allows them to keep track of every call you make, and charge you for it. But switching every connection is expensive. (It takes lots of equipment, etc.).

Data communications don't need to be switched - they carry their own destination with them, as part of their signal. So, they can function better over a cheaper network.

From this explanation, you can see the position that telephone companies are in - if the government stops enforcing their monopolies and data communication takes over, their primary market will be people who cannot afford computers. Yes, they have enormous resources, and they are capable of doing almost anything they want - so, they certainly could redesign their systems and function well in the 21st century. But at this point, it doesn't look like they are committed to that course of action, and they are not capable of changing horses quickly.

What does this mean to the electrical contractor?

It means several things:
1. The amount of data communications that our customers do is going nowhere but up. This is a big business now, but it is nothing like what it will be once the monopolies are no longer enforced.
2. While the players in this market may change, the technologies now in place will be used for a long time. The way communications signals are sent now will certainly modify over time, it will do so incrementally, no matter who owns the wires.
3. Eventually optical fiber will be installed network-wide. But this will also come incrementally, since there is so much copper already in place, and the people who own it will use it as long as they can.
4. Huge opportunities will be afoot. When markets change on a large scale, gaps open up in them. The people who can see and fill them appropriately make huge amounts of money.

Understanding Telephone Networks

Although telephones and telephone company practices may vary dramatically from one locality to another, the basic principles underlying the way they work are the same everywhere.

Every telephone consists of three separate subassemblies, each capable of independent operation. These assemblies are the speech network, the dialing mechanism, and the ringer. Together, these parts - as well as any additional devices such as modems, dialers, and answering machines - are connected together, and transfer voice communications through telephone circuits.

A telephone is usually connected to the telephone exchange by an average of about three miles of No.22 (AWG) or 0.5 mm copper wires. This wiring back to the phone company office is commonly called as *the loop*, or the *neighborhood loop*. Although copper is a good conductor, it does have resistance, especially in small gauges. The resistance of No.22 twisted pair AWG wire is 16.46 Ohms per thousand feet at 77 degrees F (25 degrees C). Telephone companies describe loop length in *kilofeet* (thousands of feet).

Because telephone apparatus is generally considered to be current-driven, all phone measurements refer to current consumption, and not voltage. The length of the wire connecting the subscriber to the telephone

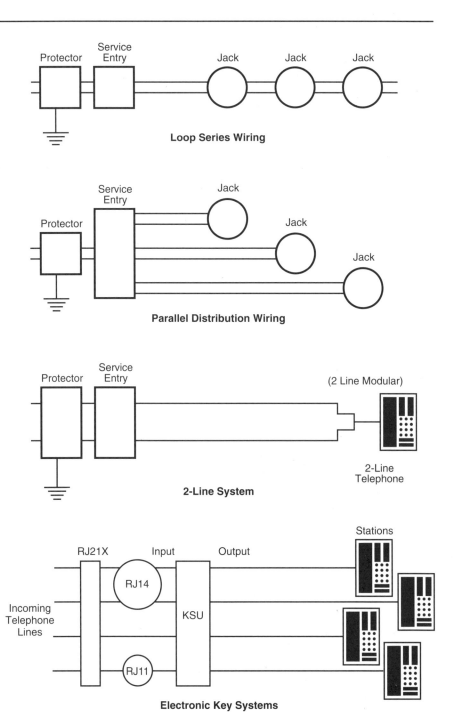

Fig. 7-1. Typical telephone system wiring methods.

exchange affects the total amount of current that can be drawn by anything attached at the subscriber's end of the line.

In the United States, the voltage applied to the line to drive the telephone is 48 VDC. In other places, 50 VDC is sometimes used. Note that telephones are peculiar in that the signal line is also the power supply line. The voltage is supplied by lead-acid cells, thus assuring a hum-free supply and complete independence from the electric company, which is why telephones work during power outages.

At the telephone exchange the DC voltage and audio signal are separated by directing the audio signal through two 2 uF capacitors and blocking the audio from the power supply with a 5-Henry *choke* (inductive coil) in each line. Usually these two chokes are the windings of a relay that switches your phone line at the exchange; in the United States, this relay is known as the "A" relay. The resistance of each of these chokes is 200 Ohms.

We can find out how well a phone line is operating by using Ohm's law and an ammeter. The DC resistance of any device attached to the phone line is often quoted in telephone company specifications as 200 Ohms; this will vary in practice from between 150 to 1,000 Ohms. You can measure the DC resistance of your phone with an Ohmmeter. Note that this is DC resistance, not impedance.

Using these figures you can estimate the distance between your telephone and the telephone exchange. In the United States, the telephone company guarantees you no lower current than 20 mA - or what is known to your phone company as a *long loop*. A *short loop* will draw 50 to 70 mA, and an average loop, about 35 mA. (Remember, in telephone work, current measurements are used, not voltage measurements.) Some exchanges will consider your phone in use and feed dial tone down the line with currents as low as 8 mA, although the telephone may operate correctly.

Telephone companies, to avoid people using their 50 volt DC feed into the home or office for other things, like the DC resistance of their lines to be about 10 megOhms when no apparatus is in use. (The state on not being in use is called *on-hook* in telephone jargon.) A phone that is on-hook can draw no more than 5 microamperes.

The phone line has an impedance composed of distributed resistance, capacitance, and inductance. The impedance will vary according to the length of the loop, the type of insulation of the wire, and whether the wire is aerial cable, buried cable, or bare parallel wires strung on telephone poles. For calculation and specification purposes, the impedance is normally assumed to be 600 to 900 Ohms. If the instrument attached to the phone line should be of the wrong impedance, you would get a mismatch, or what telephone company personnel refer to as *return loss*. A mismatch on telephone lines results in echo and whistling, which the phone company calls *singing*. You may have heard this effect on inexpensive telephones. A mismatched device can be matched to the phone line by placing resistors in parallel or series with the line to bring the impedance of the device to within the desired limits. This will cause some signal loss, of course, but will make the device usable.

A phone line is a balanced feed, with each side equally balanced to ground. Any imbalance will introduce hum and noise to the phone line and increase susceptibility to RFI *(radio frequency interference)*.

The balance of the phone line is known to your telephone company as *longitudinal balance*. If both impedance match and balance to ground are kept in mind, any device attached to the phone line will perform well, just as the correct matching of transmission lines and devices will ensure good performance in radio practice.

For older systems, the two phone wires connected to your telephone should be red and green. The red wire is negative and the green wire is positive. Your telephone company calls the green wire *Tip* and the red wire *Ring*. Most installations have another pair of wires, yellow and black, which can be used for many different purposes, if they are used at all. If two separate phone lines are installed in a home or office with older cables, you will find the yellow and black pair carrying the second telephone line. In this case, black is *Tip,* and yellow is *Ring*. Some party lines use the yellow wire as a ground; sometimes there's 6.8 volts AC on this pair to light the dials of Princess-type phones.

The above description applies to a standard line with a DC connection between your end of the line and the telephone exchange. Most phone lines in the world are of this type, known as a *metallic line.* In a metallic line, there may or may not be inductive devices placed in the line to alter the frequency response of the line; the devices used to do this are called *loading coils*.

Very long lines may have amplifiers, sometimes called *loop extenders* on them. Some telephone companies use a system called a *subscriber carrier*, which is basically an RF *(radio frequency)* system in which your telephone signal is raised up to around 100 Khz and then sent along another subscriber's pair of conductors.

When a telephone is taken off the hook, the line voltage drops from 48 Volts to between 9 and 3 Volts, depending on the length of the loop. If another telephone in parallel is taken off the hook, the current consumption of the line will remain the same and the voltage across the terminals of both telephones will drop. Bell Telephone specifications state that three telephones should work in parallel on a 20 mA loop; transistorized phones tend not to pass this test, although some manufacturers use ICs (integrated circuits) that will pass.

While low levels of audio may be difficult to hear, overly loud audio can be painful. Consequently, a well designed telephone will automatically adjust its transmit and receive levels to allow for the attenuation - or lack of it - caused by the length of the loop. This adjustment is called *loop compensation*. Telephone manufacturers achieve this compensation with silicon carbide varistors that consume any excess current from a short loop. Although some telephones using integrated circuits have built-in loop compensation, many do not; the latter have been designed to provide adequate volume on the average loop, which means that they provide low volume on long loops, and are too loud on short loops. Because a telephone is a duplex device, both transmitting and receiving on the same pair of wires, the network must ensure that not too much of the caller's voice is fed back into his or her receiver. This function, called *sidetone*, is achieved by phasing the signal so that some cancellation occurs in the speech network before the signal is fed to the receiver. Callers faced with no sidetone at all will consider the phone dead. Too much sidetone causes callers to lower their voices and not be heard well at the other end of the line. Too little sidetone will convince callers that they're not being heard.

A telephone on a short loop with no loop compensation will appear to have too much sidetone. In this case, the percentage of sidetone is the same, but as the overall level is higher, the sidetone level will also be higher.

There are two types of dialing. The old method is called pulse, loop disconnect, or rotary, and it's been with us since the 1920's. The newer dialing method is called *Touch-tone, Dual Tone Multi-Frequency* (DTMF).

Pulse dialing is traditionally accomplished with a rotary dial, which is a speed governed wheel with a cam that opens and closes a switch in

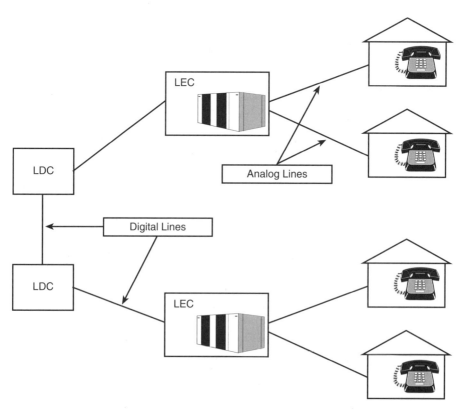

Fig. 7-2 Traditional telephone network (POTS).

series with your phone and the line. It works by actually disconnecting or *hanging up* the telephone at specific intervals. The United States standard is one disconnect per digit, so if you dial a 1, your telephone is disconnected once. Dial a seven and you'll be momentarily disconnected seven times; dial a zero, and you'll *hang up* ten times.

Although most exchanges are quite happy with rates of 6 to 15 Pulses Per Second (PPS), the more accepted standard is 8 to 10 PPS. Some modern digital exchanges, free of the mechanical inertia problems of older systems, will accept a PPS rate as high as 20.

Touch tone, the modern form of dialing, is faster and less prone to error than pulse dialing. Compared to pulse, its major advantage is that its audio band signals can travel down phone lines further than pulse, which can travel only as far as your local exchange. Touch-tone can therefore send signals around the world via telephone lines, and can be used to control phone answering machines and computers.

Bell Labs developed DTMF to have a dialing system that could travel across microwave links and work rapidly with computer controlled exchanges. Each transmitted digit consists of two separate audio tones that are mixed together. The four vertical columns on the keypad are known as the high group and the four horizontal rows as the low group; the digit 8 is composed of 1336 Hz and 852 Hz. The level of each tone is within 3 dB of the other, (the telephone company calls this *Twist*). A complete touch-tone pad has 16 digits, as opposed to ten on a pulse dial. Besides the numerals 0 to 9, a DTMF "dial" has *, #, A, B, C, and D. Although the letters are not normally found on consumer telephones, the IC chip in the phone is capable of generating them.

Data Over the Telephone Lines

Sending data over phone lines comes in two basic forms:
1. Using the phone company's switched network for low-speed modem traffic.
2. Using dedicated circuits that can function at high speeds and more or less independently of the network switching system.

Modem Traffic

The use of *modems* (modulator-demodulators) is very widespread, and requires only standard telephone lines. In the most cases, you can connect a modem to the end of an existing telephone line, and send up to 56,000 bits per second over the lines; and while a 56 K modem is a lot faster than the 2400 baud modems a lot of us started with, it is still a small fraction of the signal required for even one channel of television.

T1 Lines

At the present moment, if you want anything even resembling real bandwidth from the telephone company, you'll have to get a dedicated T1 line. This does not give you a high-speed connection from your home or office to your choice of locations - it only gives you high-speed access from your home or office to one specific location. Usually this location is a place where you can connect to an Internet backbone (main line), and send routed signals through the Internet. (*Routed* signals being the type that contain their own destination address as part of the message, and do not go through the telephone company's switches.) You do not pay per-call for T1 lines - you pay a flat (high) monthly fee.

When you lease a T-1 line from either the telephone company or an alternative service provider (there are a few Internet companies that buy T1 lines in bulk from the phone companies, and resell them at lower prices than the phone companies charge), you receive access to one of the company's connectors, to which your network attaches. (In telephone company jargon everything beyond the connector is called *customer premises equipment*, or CPE.)

However, you can't attach your router to the *telco* (telephone company) connector by simply plugging in a cable. To make sure that a network interfaces properly with a telephone line, several functions must be performed. *CSUs* and *DSUs* are the devices that are used to handle these functions.

The CSU

A *channel service unit*, or CSU, is the first device the external telephone line encounters on the customer premises. In the early 1980s, CSUs were always owned by the telephone company, which leased the devices to customers. But during the course of telecommunications deregulation, users were allowed to buy and install their own CSUs.

One of the principal functions of a CSU is to protect the carrier and its customers from any weird events your network might introduce onto the carrier's system.

A CSU provides termination for the telephone line and performs line conditioning and equalization. It also supports *loopback tests* for the carrier, which means that the CSU can reflect a diagnostic signal to the telephone company without sending it through any customer premises equipment. This is done for diagnostic purposes. CSUs often have indicator lights or LEDs that identify lost local lines, lost telco connections, and loopback operation.

When CSUs were provided only by the telephone company, they were generally powered by the telephone line itself. Now, CSUs generally need their own power supply, and perhaps a backup power supply at the user site.

The DSU's Duty

Data service units - sometimes referred to as *digital service units* - or DSUs, are connected between a CSU and customer equipment such as routers, multiplexers, and terminal servers. DSUs are commonly equipped with RS-232 or V.35 interfaces. Their main function is to adapt the digital data stream produced by the customer equipment to the signaling standards of the telephone carrier equipment, and vice versa.

The digital signals produced by customer devices with throughput less than 56Kbps are *asynchronous*, which means that each byte is distinguished by start and stop bits and that the time interval between bytes varies. However, most customer devices in the telecommunication system use *synchronous* signaling, in which senders and receivers coordinate with timing clocks to identify the boundaries between units of data.

In these cases, the DSU may be called upon to parcel out incoming asynchronous data at the stable rate the carrier line expects, and to wrap start and stop bits around incoming synchronous data before passing it along to the user network.

Computer Telephony Terms

Following are some of the key terms that are used for defining links between data communications and telecommunications:

ADSL - Asymmetric Digital Subscriber Line. A method of carrying high-speed traffic over existing copper twisted-pair wires. Currently in the trial phase, ADSL offers three channels: a high-speed (between 1.5Mbps and 6.1Mbps) downlink from the carrier to the customer, a full-duplex data channel at 576Kbps, and a plain old telephone service (POTS) channel. A key feature of ADSL is that POTS is available even if the extra ADSL services fail.

ANI - Automatic Number Identification. A system that identifies the telephone number of the calling party for the call recipient, is known to most consumers as caller ID. When using a T-1 line, the ANI information also includes the geographic coordinates of the originating call's central office.

DS-0 A Digital Signal Level Zero. This is one of the 64,000bps circuits in a T-1 or E-1 line. It consists of 8-bit frames transmitted at 8,000 frames per second. The usable bit rate is often only 56,000bps. DS-1, or Digital Signal Level One, is often used as a synonym for T-1, but it more precisely refers to the signaling and framing specifications of a T-1 line. See T-1.

DTMF - Dual-tone Multifrequency. A description of the audio bleeps you hear when you dial a Touch-Tone telephone. Each row and each column of keys is assigned a separate frequency. Each key's frequency is produced by combining row and column frequencies. For example, the second column's assigned frequency is 1,336Hz, and the third row has 852Hz; pressing the number 7 generates both those tones. By decoding the two frequencies, the telephone company's central office, your PBX, or an interactive voice response system can detect which button was pressed. ("Touch-Tone" is AT&T's trademark for DTMF.)

Frame relay. This refers to a shared-bandwidth wide area network based on a subset of High-level Data-link Control (HDLC) called LAP-D (link access procedure-D channel). Frame relay is designed to be carried over high-speed, high-accuracy links such as T-1 or the still emerging T-3; a 56kbps line is the most common implementation. Individual frames can vary in size, but they are usually 4,096 bytes. Users reserve a specific data rate called the CIR, or committed information rate, but users can attempt to burst data at higher rates. Extra frames are discarded if the carrier's network doesn't have sufficient capacity.

HDSL - The High bit-rate Digital Subscriber Line. This circuit offers a full-duplex 784kbps connection over two twisted pairs. HDSL can carry either a full T-1 connection over two twisted pairs or a fractional T-1 connection over a single twisted pair of wires.

ISDN Integrated Services Digital Network. A telephone system service that provides access to both the public-switched telephone network and to packet-switched services (such as X.25 and frame relay). ISDN offers two types of channels: B (bearer), which are 64kbps voice

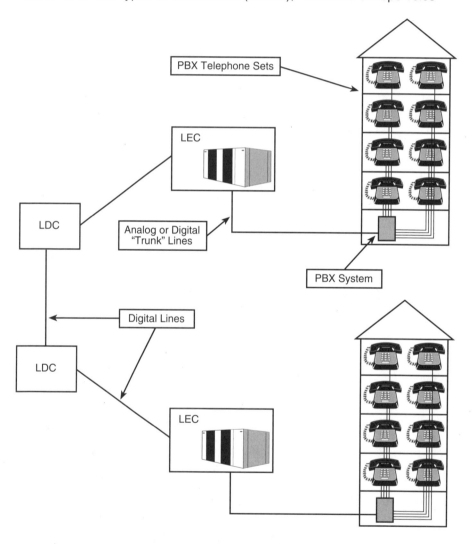

Fig. 7-3 PBX (Private Branch Exchange).

channels, and D (delta), which are channels for setup, coordination, and control. Telephone companies offer ISDN in two main varieties: basic rate interface (BRI), which contains two B channels and an 8kbps D channel, and primary rate interface (PRI), which has 23 B channels and a 64kbps D channel.

IVR - Interactive Voice Response. The basic voice-mail system that can decode DTMF signals, which is how your call can be routed without the aid of a real human being.

PBX - Private Branch Exchange. A telephone switch used within a business or other enterprise-as opposed to the switches used at a public telecommunication service provider's central office (CO). PBXs might offer basic telephone service, some level of computer-telephony integration (CTI), voice mail, or other features. When you dial within your company, the PBX provides the dial tone; when you dial 8 or 9 for an outside line, the CO provides the dial tone.

T-1 T-1 is a North American standard for point-to-point digital circuits over two twisted pairs. A T-1 line carries 24 64,000bps channels (also known as DS-0) for a total usable bit rate of 1,536,000bps (if you include extra bits used to synchronize the frames, the actual bandwidth is 1,544,000bps). Customers may lease a fractional T-1, using only some of the 24 T-1 slots. A T-1C contains two T-1 lines; T-2 supports four T-1 circuits. A T-3 communications circuit supports 28 T-1 circuits, and a T-4 consists of 168 T-1 circuits. E-1 through E-5 are similar standards used in Europe and Japan, but they offer different numbers of channels.

TAPI Telephony Application Programming Interface is a method promoted by Microsoft and Intel for letting PCs control telephones. TAPI applications can dial a telephone from within software and check caller ID, as well as perform other functions. TAPI links PC workstations and individual telephones, as opposed to TSAPI (see below), which links the PC server to the PBX.

TSAPI - Telephony Services Application Programming Interface. A system promoted by Novell and AT&T to integrate PCs and PBX servers. TSAPI allows computer control of most aspects of the local telephone system. Contrast this to TAPI, which links the PC and the local telephone sitting on the user's desk.

X.25 This term refers to a standard for packet-based wide area networks. For both leased lines, such as T-1s, and public-switched connections, like ISDN's B-channel links, you pay by the minute or month. However, an X.25 connection has the advantage of being measured and billed by the number of packets or bytes actually sent or received.

The Operation of Telephones

Modern telephones operate on essentially the same principles that were developed over 100 years ago. They use a single pair of wires that connect the phones and a power source. When phones are connected, the power source causes a current to flow in a loop, which is modulated by the voice signal from the microphone in one handset and excites the earphone in the opposite headset.

Dialing was originally done by a rotary dial that simply switched the current on and off in a number of pulses corresponding to the number dialed. Dialing is now mostly (but not entirely) accomplished with tones.

By operating on a current loop, phones can be powered from a central source and extended simply by adding more wire and phones in parallel. Now, most phones are electronic (that is, they use semiconductors rather than electromechanical devices), but they use the same type of wiring, frequently called *current-loop wiring*.

Telephone wiring is simple because the *bandwidth* (bandwidth is similar to speed, a low bandwidth requires lower frequencies, and a higher bandwidth requires a higher frequencies) of telephone signals is low, generally around 3,000 *hertz* (cycles per second.) (Computer modems use sophisticated techniques to send much faster digital signals over low-bandwidth phone connections.)

Because of the low bandwidth and current loop transmission, telephone wire is easy to install and test. It can be pulled without fear and if it is continuous, it should work.

The most common pieces of *telecom* equipment that you will be mounting are the following:

Telephones. The main concern here is that the proper phone is connected in the proper location.

Punch down blocks. It is important that punch down blocks are mounted securely. If not, connections will be difficult to make, and the block will loosen during the termination process.

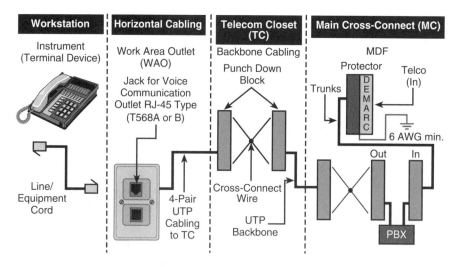

Fig. 7-4. Shown is the layout of a basic business telephone system. The protector (righthand side) is a surge suppressor, usually the metal oxide varistor (MOV) type. The box marked "Demarc" is the point of demarcation between the telephone company's and customer-owned equipment.

Wall jacks and connection modules. Here there are two concerns. First, that the conductors are terminated in the proper order, and second, that the jack or module is secure; it may suffer a fair amount of abuse during its useful life.

Equipment racks. Larger systems may require equipment racks. These

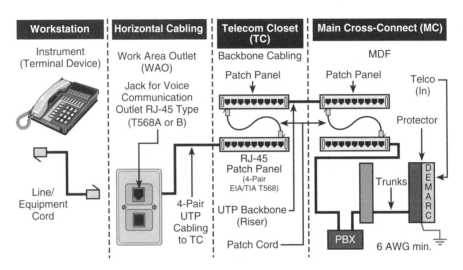

Fig. 7-5. Shown is a very similar system to that of the figure above, but designed with patchpanels rather than punch-down blocks. Either design is fine from a performance standpoint. However, patchpanels are generally more expensive and require you to terminate each cable with a modular plug before plugging them into the patchpanel.

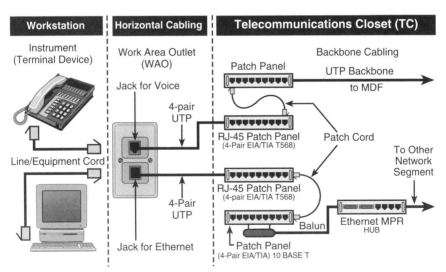

Fig. 7-6. Shown is a very common combination of voice and system: voice and Ethernet.

101

must be not only secure, but must also be installed plumb and level.

Central consoles. Central consoles must be properly placed, and must be installed near a source of electrical power.

Computer cards. Especially when modifying an existing system, you will have to install computer cards in existing personal computers. This requires some experience with computers, and just a bit of gentleness.

Large central units. Large central units are essentially dedicated computers connected to switching units. When installing these, you have several concerns. First of all, you must be sure that there is plenty of space to install the unit and to access it during the installation process and for servicing afterward. You must also be sure that unit is properly supplied with power and properly grounded

Digital Telephony

At the beginning of the telephone business, all telephone circuits were analog. They were designed around the characteristics and problems of analog current, and were maximized for that use. Since computers have arrived, however, a great number of people want to send digital signals over telephone lines. Both recently and from now on, growth in communications will be overwhelmingly digital. Analog is on its way out.

Understanding Analog and Digital Lines

It is important to understand the differences between analog and digital circuits. Analog signals vary continuously, and they represent particular values, such as the volume and pitch of a voice or the color and brightness of a section of an image. Digital signals have meaning only at *discrete* levels, that is, the signal is either on or off, present or absent, 1 or 0.

Analog telephone lines are the *legacy* (old, traditional) systems of the telephone industry. Nearly all residential telephone lines are analog. Fifty-year-old telephones will probably work on your local loop (the connection between your home telephone jack and the telephone company's *central office*). The central office is usually not a large downtown building, since the average local loop is about 2.5 miles, so the "central office" is most often an inconspicuous building in or near your neighborhood.

When you talk on the telephone, the microphone in the "receiver" (when you are talking, you are actually using it as a transmitter) produces an analog signal that travels to the central office and is switched either to another local destination or to other switching offices that connect it to a remote destination. Dialing the telephone produces the in-band signals that tell the switching system where to route the call.

The telephone companies have learned a great deal about the electrical characteristics of human voice signals over the years, and they have determined that we will be reasonably satisfied with voice signals that do not transmit frequencies below 300Hz or above 3,100Hz. Note that high fidelity is usually considered to be a system that can reproduce frequencies between 20Hz and 20KHz without distortion-while voices are recognizable with the standard telephone frequency range, that range of frequencies is likely to be inadequate for other types of sounds-for instance, music sounds lousy over the telephone. Telephone companies allow an analog telephone channel a bandwidth of 4,000Hz to work with.

(Some urban business telephone users have digital services direct to their PBXs or data communication devices, or digital ISDN lines.)

At the central office, the odds are that the analog signal will be digitized to be switched across the telephone network. Aside from a few remote areas, the U.S. telephone network that interconnects central offices uses digital signaling.

The local loop is sometimes referred to as "the last mile," because residences, generally saddled with analog-only transmission facilities, are rarely capable of bandwidth greater than 4,000Hz.

Bandwidth and the 4kHz Channel

Bandwidth determines the speed at which a signal can be transmitted. Think of bandwidth as a road or freeway. The larger the road and less congested, the faster you can cruise on it. As it applies to computer communications, the information you receive on your computer screen is the data that travels on the bandwidth. Your modem converts the analog signal into a digital signal and vice versa. If you are just using your phone line, an analog connection called *POTS* (plain old telephone service), the fastest you can normally go is between 28.8kbps and 56kbps. Think of this phone line as a dirt paved road to the Digital Superhighway.

The dirt paved road is just a phone line from your local company (POTS - Plain Old Telephone Service). This is an analog service that allows you to connect to the interconnected fiber optic backbones that make up the long distance and Internet systems, nicknamed *the Cloud*. This is the long distance network which is basically 100% digital in the United States.

The national telephone systems are very rarely digital all the way to the curb.

Modems convert digital signals from a computer into analog signals in the telephone frequency range. There is a hard upper limit to the capacity of a channel with a given bandwidth. A channel's throughput in bits per second depends on the bandwidth and the achievable signal-to-noise ratio. The current top throughput rate for modems of 56kbps is right at the limit, and frequently cannot run at full speed. As users of even 28.8kbps modems know, the actual throughput achievable on normally noisy analog lines is rarely the full-rated value, and it may be much lower. Compression and caching and other tricks can mask the limit to an extent.

The modem was certainly a big breakthrough in technology. It allowed computers to communicate with each other by converting their digital communications into an analog format to travel through the public phone network. But there is a limit to the amount of information that a common analog telephone line can carry. Currently, it is about 56 kbps.

When the telephone company reverses the process and digitizes an analog signal, it uses a 64kbps channel. (This conversion is a worldwide standard.) One of these channels, called a *DS0* (digital signal, level zero), is the basic building block for telephone processes. You can combine (the precise term is *multiplex*) 24 DS0s into a *DS1*. If you lease a T-1 line, you get a DS1 channel. With synchronization bits after each 192 bits (that is, 8,000 times a second), the DS1 capacity is 1.544Mbps (the product of 24 and 64,000, with another 8,000 sync-bits added).

Dedicated Lines, Switched Lines

The second important distinction to make about telephone lines is whether they are dedicated circuits you lease or switched services you buy. If you order a T-1 line or a low data-rate leased line such as *dataphone digital service* (DDS), you are renting a point-to-point circuit from the telephone company. You have dedicated use of such a circuit-with 1.544Mbps (T-1) or 56Kbps (low data rate) of capacity, respectively.

These services are sold as permanent virtual circuits, where the customer specifies the end points. Switched services, such as residential analog telephone service, are services purchased from the telephone company. You can select any destination on the telephone network and connect to it through the network of public switches. You generally pay for connect time or actual traffic volume, so unlike a dedicated line, the bill will be low if usage is low. Switched digital services include X.25, Switched 56, ISDN Basic Rate Interface (BRI), ISDN Primary Rate Interface (PRI), Switched Multimegabit Data Service (SMDS), and ATM. It's also possible to set up private networks that supply these services using your own switching equipment and leased lines or even privately owned lines-for example, if you are a university, a railroad, or a municipal utility.

If the circuit provided by the phone company is already a digital circuit, there is no need for a modem to provide digital-to-analog conversion services between the *terminal equipment* (equipment such as computers, fax machines, and digital telephone instruments) and the telephone system. Nonetheless, customer premises equipment must present the correct electrical termination to the local loop, transmit traffic properly, and support phone company diagnostic procedures.

A line that supports ISDN BRI service must be connected to a device called an *NT1* (network termination 1). In addition to the line termination and diagnostic functions, the NT1 interface converts the two-wire local loop to the four-wire system used by digital terminal equipment. For digital leased lines, you'll need T-1 and DDS-and for the digital services, the digital subscriber line from the phone company needs to be terminated by a channel service unit, or CSU. The CSU terminates and conditions the line and responds to diagnostic commands. Customer terminal equipment is designed to interface with a data service unit (DSU), which hands over properly formatted digital signals to the CSU. CSUs and DSUs are often combined into a single unit called (rather obviously) a *CSU/DSU*. The DSU may be built into a router or multiplexer. So even though end-to-end digital services don't require modems, a piece or two of interfacing hardware is always required for connectivity.

Phone Services

While 33.6 kbps or 56 kbps is a stretch for most local loops configured for analog service, the same twisted-pair wiring running between a house and a central office is probably capable of supplying ISDN BRI service, with 128kbps of data throughput capacity and another 16kbps of control and setup capacity.

Here's how this is possible: Analog telephone circuits are heavily filtered to keep the signals attenuated outside their 4kHz bandwidth. Digital circuits don't need to be filtered the same way, so the twisted pair cable can support a much greater bandwidth, which allows greater throughput.

Leased 56kbps and 64kbps lines and services that run on these lines, such as frame relay and Switched 56, may be delivered on a two-wire digital line or on a four-wire digital line (which has two separate pairs - one pair for transmitting, and one pair for receiving). T-1 lines as well as ISDN PRI and frame relay are often delivered on four-wire digital lines or perhaps on optical fiber. T-3 lines are sometimes coaxial cable, but most high-capacity traffic is carried on optical fiber. While ISDN is getting a lot of attention as a high-capacity, wide-area connection, there are other systems competing to be the residential service for the "last mile." Several companies (PairGain, AT&T, Paradyne) are promoting the *high bit-rate digital subscriber loop* (HDSL) technology. These products serve to equalize local loops dynamically, making it possible to support DS1 throughput - 1.544Mbps - over most existing twisted pair local loops, provided that new HDSL devices (relatively small pieces of electronic equipment) are installed at both ends.

With standard 24-gauge wire, HDSL can be used successfully on local loops up to 2.3 miles long, with no repeaters. Ordinary T-1 circuits require repeaters at least every 3,000 feet to 5,000 feet. If you want to transport DS1 levels of traffic over the last mile, the alternatives to HDSL are to install "fiber to the curb" at great expense or to install several repeaters on each line, which is not as expensive as all-new fiber but is still costly and imposes a large maintenance cost on the telephone company (and, ultimately, on the customer).

HDSL is not even the final word on improving the throughput of the last mile. *Asymmetrical digital subscriber line technology* (ADSL), an extension to HDSL, is expected to support throughput as high as 6Mbps in a single direction, with a much lower throughput (perhaps 64kbps-in the other direction). In an open market where customers pay for a telephone service based on the actual cost of delivering that service (as opposed to rates set in Congress), a high percentage of analog telephone customers could receive ISDN PRI (or another T-1 service) at a price comparable to today's cost for ISDN BRI. In most cases, telephone companies can install ISDN line equalizers and keep the savings to themselves. As is so often the case with legislated services, there's no requirement that the rates be rational.

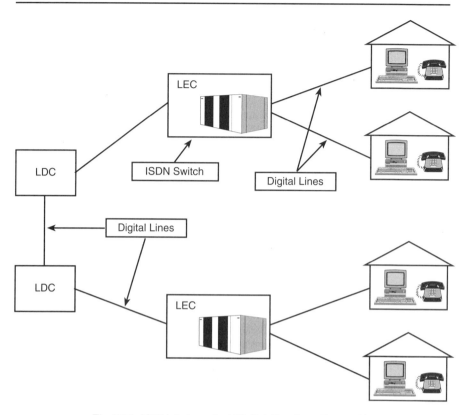

Fig. 7-7. ISDN (Integrated Digital Services Network).

ISDN

Currently, your commercial customers are using T1 lines for their communication needs. But residential and small business customers are interested in ISDN lines.

As we have previously stated, ISDN (integrated services digital network) is a system of digitizing phone networks that has been in the works for over a decade. This system allows audio, video, and text data to be transmitted simultaneously across the world using end-to-end digital connectivity.

You should know that there have been a lot of problems with ISDN lines, resulting in a lot of jokes like the following:

Q. What does ISDN stand for?
A. It Still Does Nothing.

Telephone companies have not handled ISDN very well. This is understandable, considering the deep entrenchment of the phone companies, and how slowly they adapt to new technologies. With ISDN being the first digital residential service, a lot of problems showed up. (Think about this when recommending ISDN to your customers - you might want to check on your local phone company first.)

ISDN allows multiple digital channels to be operated simultaneously through the same regular phone jack in a home or office. The change comes about when the telephone company's switches are upgraded to handle digital calls. Therefore, the same wiring can be used, but a different type of signal is transmitted across the line.

Previously, it was necessary to have a phone line for each device you wished to use simultaneously. For example, one line each for the phone, fax, computer, and live video conference. Transferring a file to someone while talking on the phone, and seeing their live picture on a video screen would require several expensive phone lines.

Using *multiplexing* (a method of combining separate data signals together on one channel such that they may be decoded again at the destination), it is possible to combine many different digital data sources and have the information routed to the proper destination. Since the line is digital, it is easier to keep the noise and interference out while combining these signals.

ISDN technically refers to a specific set of services provided through a limited and standardized set of interfaces. This *architecture* (design of the system) provides a number of integrated services currently provided by separate networks.

ISDN adds capabilities not found in standard phone service. The main feature is that instead of the phone company sending a ring voltage signal to ring the bell in your phone, it sends a digital package that tells who is calling (if available), what type of call it is (data/voice), and what number was dialed (if multiple numbers are used for a single line). ISDN phone equipment is then capable of making intelligent decisions on how to answer the call. In the case of a data call, baud rate and protocol information is also sent, making the connection instantaneous.

One of the most promising technologies for breaking the bandwidth bottleneck is Asymmetric Digital Subscriber Line (ADSL, sometimes called *Asynchronous* Digital Subscriber Line). ADSL is one of several similar technologies generically called *xDSL*. Another popular variant is HDSL—High bit-rate Digital Subscriber Loop.

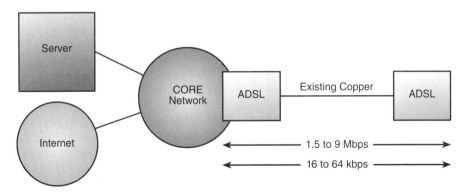

Fig. 7-8. *The basic ADSL connection. Note that the speed upstream is much slower than downstream.*

All the xDSL technologies involve the installation of electronic boxes on the ends of relatively standard telephone lines, which allows for transmission speeds of as much as 1.5 megabits/sec for HDSL and 6 megabits/sec for ADSL.

ADSL is really an extension of HDSL. It's not only the fastest of the xDSL technologies, but it also handles packet-switched transmission technologies, such as ATM (Asynchronous Transfer Mode). Sending data in self-addressed packets, via routed rather than switched circuits, is a cheaper way to send all forms of data than via voice circuits. This is how information is transmitted over the Internet, and will eventually become the standard for all transmissions.

ADSL Technology

ADSL uses three transmission channels:
1. A high-speed (between 1.5 Mbps and 6.1 Mbps) downlink from the carrier to the customer.
2. A full-duplex data channel at 576 kbps.
3. A plain old telephone service (POTS) channel.

A key feature of ADSL is: POTS remains functional even if the other ADSL channels drop out.

In the future, ADSL should support *throughput* (the amount of signal, measured in megabits, put through the circuit) as high as 6 Mbps in a single direction, with a much lower throughput of about 64 kbps in the other direction.

An ADSL circuit connects ADSL modems on both ends of a twisted-pair line. These modems create three information channels: A high-speed downstream channel, medium speed duplex channel, and POTS or ISDN channel. The POTS/ISDN channel splits off from the digital modem by filters, guaranteeing uninterrupted POTS/ISDN, even if ADSL fails. The high-speed channel ranges from 1.5 Mbps to 6.1 Mbps, while duplex rates range from 16 kbps to 640 kbps. You can submultiplex each channel to form multiple, lower-rate channels, depending on the system.

ADSL modems provide data rates consistent with North American and European digital hierarchies. With various speed ranges and capabilities, the minimum configuration provides 1.5 Mbps or 2.0 Mbps downstream and a 16 kbps duplex channel; others provide rates of 6.1 Mbps and 64 kbps duplex. Products with downstream rates up to 8 Mbps and duplex rates up to 640 kbps are available today. ADSL modems will carry *ATM* transmissions with variable rates.

Downstream Bearer Channels		Duplex Bearer Channels	
n × 1.536	1.536 Mbps	C Channel	16 kbps
	3.072 Mbps		64 kbps
	4.608 Mbps	Optical Channels	160 kbps
	6.144 Mbps		384 kpbs
n × 2.48 Mbps	2.048 Mbps		544 kbps
	4.096 Mpbs		576 kbps

Fig. 7-9. Technical specifications of ADSL channels.

Downstream data rates for ADSL depend on a number of factors, including the length of the copper line, its wire gauge, presence of bridged taps, and cross-coupled interference. Line attenuation increases with line length and frequency, and decreases as wire diameter increases. Ignoring bridged taps, ADSL will perform as follows:

Data Rate	Wire Guage	Distance	Wire Size	Distance
1.5 or 2 Mbps	24 AWG	18,000 ft	0.5 mm	5.5 km
1.5 or 2 Mbps	26 AWG	15,000 ft	0.4 mm	4.6 km
6.1 Mbps	24 AWG	12,000 ft	0.5 mm	3.7 km
6.1 Mbps	26 AWG	9,000 ft	0.4 mm	2.7 km

While distances vary from one *telco* (local telephone company) to another, ADSL (based upon the chart shown above) can cover up to 95% of a local loop, depending on the desired data rate. Reaching customers beyond these distances is possible with fiber-based *DLC* (digital loop carrier) systems. As these systems become commercially available, telephone companies can offer access in a relatively short time.

ADSL modems use a special type of error correction ("forward error correction") that dramatically reduces the errors caused by impulse noise. This is necessary for the successful transmission of digital compressed video, which may be one of the more popular transmissions over ADSL circuits.

ADSL depends upon digital signal processing and algorithms (structured mathematical formulas) to get large quantities of information through twisted-pair telephone lines. Special transformers, analog filters, and A/D converters are also factors.

Long telephone lines may attenuate signals at 1 MHz (the outer edge of the band used by ADSL) by as much as 90 dB, forcing analog sections of ADSL modems to work very hard to support large dynamic ranges, separate channels, and maintain low noise figures.

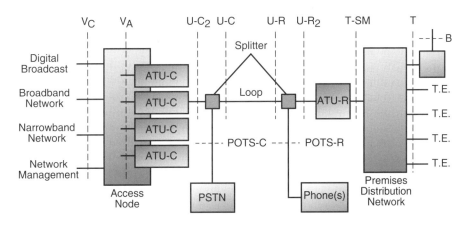

Fig. 7-10. Detailed block diagram of complex ADSL network.

Tech Details

Here are some technical details for added interest:

To create multiple channels, ADSL modems divide the available bandwidth of a telephone line in two ways—Frequency Division Multiplexing (FDM) and Echo Cancellation. FDM assigns one band for upstream data and another for downstream data. The downstream path is then divided by time division multiplexing into one or more high-speed channels and one or more low-speed channels. The upstream path is also multiplexed into corresponding low-speed channels.

Echo cancellation assigns the upstream band to overlap the downstream, separating the two by means of local echo cancellation, a technique well-known in V.32 and V.34 modems. With either technique, ADSL splits off a 4 kHz region for POTS at the DC end.

An ADSL modem organizes the aggregate data stream created by multiplexing downstream channels, duplex channels, and maintenance channels together into blocks, then attaches an error correction code to each block. The receiver corrects errors that occur during transmission up to the limits the code implies and the block length. The unit may, at the user's option, also create "superblocks" by interleaving data within "sub-blocks." This allows the receiver to correct any combination of errors within a specific span of bits, allowing for effective transmission of data and video signals.

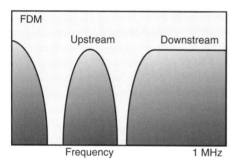

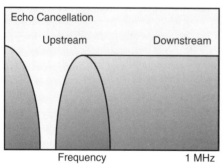

Fig. 7-11. Signal conditioning for ADSL.

The Importance of ADSL

ADSL's transmission rates of 1.5 Mbps to 6 Mbps are only a fraction of what would be available with optical fiber to the home or office (which would be at least hundreds of gigabits). Compared to regular telephone circuits, it's phenomenal.

These rates would expand existing bandwidth by a factor of 50 or more. ADSL could transform the existing information network from one limited to voice, text, and low-resolution graphics to a powerful, ubiquitous system capable of bringing multimedia (including full motion video) into millions of homes and offices.

ADSL could play a crucial role over the next 10 or more years as telephone companies enter new markets for delivering information in video and multimedia formats. Success of these new services depends on reaching as many subscribers as possible. By bringing movies, television, video catalogs, remote CD-ROMs, corporate LANs, and the Internet into homes and small businesses, ADSL could make these markets viable for telephone companies and application suppliers alike.

Will this really happen?

This is the big question. ADSL modems test successfully in more than 100 telephone companies, telecom operators, and thousands of lines installed in various technology trials in North America, Europe, and Asia. Several telephone companies plan market trials using ADSL, principally for data access, but also including video for such applications as personal shopping, interactive games, and educational programming.

Semiconductor companies introduce transceiver chipsets already used in market trials. These chipsets combine off-the-shelf components and programmable digital signal processors.

Some industry "experts" expect the phone companies to act according to market conditions to provide the bandwidth to the home desired by their customer. However, this may never happen on a large scale.

Local telephone systems are not designed for high-speed circuits like ADSL—they were built to send each call through a switching system. That's how they charge for their services. ADSL, and all new data systems, are designed for Internet-type *routed* systems. That is, they do not run through switches. Instead, installers route them from one place to another. This provides a less expensive, more versatile network. However, it leaves the phone companies with less to charge for.

Fundamentally, the choices facing phone companies are:
A. Continue to charge $1,000.00 per month for 1.5 Mb/s T1 lines, or
B. Invest in equipment to sell 6 Mbps ADSL lines for $50.00 per month.

You can see why it's not in the immediate interest of the phone companies to jump into ADSL. And since local telcos have government-enforced monopolies, there's no direct competition to spur them into action.

A few years ago, some touted ISDN as the new high-speed digital circuit that would provide high-speed service to homes and businesses. Since telcos didn't want it to succeed, it died on the vine. The same could happen to ADSL. And as the technology develops, support and obtainability won't be easy.

Computer makers and software designers have prototypes sitting on their shelves that would give you incredible features, such as scanning a video log and downloading your favorite NFL game, movie, or old TV show at a moment's notice. But these products will stay on the shelf until there is enough bandwidth to make them usable. There is no point to download the 1986 Super Bowl game if it takes four hours to do it, and it ties up your phone lines as well. But, if there is fiber to the home, you can download the whole game in the time it takes you to grab a Coke from the fridge.

The telephone company/government regulatory establishment has things closed off for the moment, and many observers wonder when they will see a need to change. Time will tell.

Other Digital Services

Switched 56. Unlike ISDN, this service is already offered by most carriers. It creates a virtual network over existing public phone lines with a 56 kbps data rate. This service is cheap, but slow; therefore, it is ideal for intermittent data swapping between WANs.

SMDS (Switched Multi-megabit Data Services). Using a connectionless networking plan, each SMDS packet has its own address and does not require a virtual circuit. Proposed speeds are from 1.5 Mbps to 45 Mbps using a fixed-length packet of 53 KB. Many regional carriers are beginning to offer this service for local traffic.

ATM (Asynchronous Transfer Mode). Using the same 53 KB packets as SMDS, ATM uses virtual circuits to transfer data at speeds of 34 Mbps to multiple gigabits per second.

Multiplexing

Multiplexing is the process of combining numerous bit streams or signals so they all can be carried efficiently. All carrier-based transmission systems use frequency division multiplexing as a way of creating discrete channels, and preventing signals on different channels from interfering with each other. Frequency division multiplexing works by assigning channels to different blocks of spectrum. A single TV channel might operate in the 6MHz block from 54 to 60 MHZ, while the next channel occupies the 60 to 66 MHZ block. The third channel might occupy the 66 to 72 MHZ space.

Baseband digital transmission systems typically use time division principles to separate different messages. TDM interleaves data packets from multiple messages, separating them slightly in time.

Optical transmission systems can use wave division multiplexing, where a number of separate optical wavelengths are offset from each other by about 1.6 nanometers, for example. Some wavelength division systems transmit 16 separate

Pair	Tip (+) Color	Ring (−)Color
1	White	Blue
2	White	Orange
3	White	Green
4	White	Brown
5	White	Slate
6	Red	Blue
7	Red	Orange
8	Red	Green
9	Red	Brown
10	Red	Slate
11	Black	Blue
12	Black	Orange
13	Black	Green
14	Black	Brown
15	Black	Slate
16	Yellow	Blue
17	Yellow	Orange
18	Yellow	Green
19	Yellow	Brown
20	Yellow	Slate
21	Violet	Blue
22	Violet	Orange
23	Violet	Green
24	Violet	Brown
25	Violet	Slate

Fig. 7-12. Standard telecom color coding.

wavelengths on a single optical fiber, stacking signals in the 1557.36 to 1545.32 nanometer range. Each of the 16 wavelengths can carry data at rates between 150 Mbps and 2.4 Gbps, for example. Newer systems can carry 32 wavelengths on a single fiber, using spacing of between

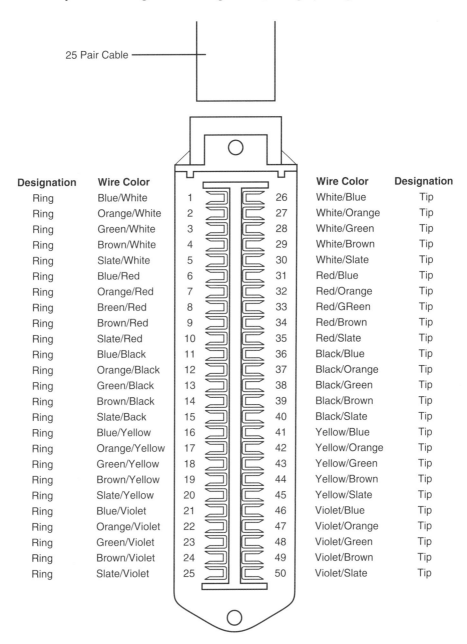

Fig. 7-13. 25-pair color coding/ISDN contact assignments.

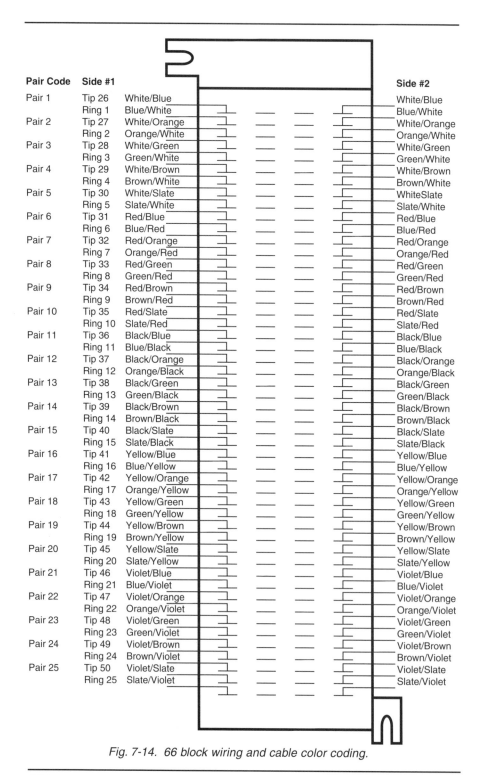

Fig. 7-14. 66 block wiring and cable color coding.

0.8 and 1.0 nanometers.

Code division multiple access (CDMA) is another multiplexing method used by cellular and PCS networks. CDMA uses frequency division to create channels, and then a digital coding technique to stack time division signals within each channel.

Space division multiplexing is another carrier-based communications networks separate channels. Cellular and PCS networks are good examples of space division, where frequencies and channels are spatially separated from each other.

Analog To Digital Conversion

Before naturally-occurring information (speech, music, text, images) can be transmitted over a digital network, the information first must be converted to digital format. It then can be modulated. Modulation is the process of impressing a high-energy waveform on the digital signal so that it can be transmitted over distance. Of the three basic processes central to digital communication systems, analog-to-digital conversion is the first step.

Analog-to-digital conversion (A/D) is the process of transforming a naturally occurring analog (wave form) signal into a digital format. The analog waveform can represent speech, television or music. The issue here is that all these types of signals are perceived by the eye and ear in analog format, despite their transmission in digital form. The A/D process involves three discrete steps:

1. Sampling
2. Quantizing
3. Coding

Sampling is the process of reducing a very complex waveform into a manageable numerical form (a string of ones and zeros) without losing the richness and nuances of the waveform. Key to the process is a mathematical proposition, developed by a Bell Laboratories engineer Harry Nyquist in the 1930s. The Nyquist sampling theorem is a fundamental concept underpinning all digital communications. The Nyquist theorem suggests that any complex waveform can be reconstructed from a limited number of discrete samples.

Sampling involves the taking of instantaneous measurements of the waveform, especially amplitude. The more samples are taken, the more faithfully the original waveform can be reconstructed. In fact, the Nyquist theorem states, all the information contained in the waveform can be recaptured when precisely twice as many samples are taken as the highest frequency of the waveform. If, for example, the highest frequency in a human voice is 4,000 cycles per second (4 kHz), then 8,000 samples must be taken each second. The objective is to measure the amplitude (height) of the waveform at set time intervals.

Once the waveform has been sampled, it is quantized. Quantizing is the process of mathematically assigning a value to each of the discrete samples. The basic concept is that of "rounding" values, much as all numbers on a scale of one to ten could be rounded to the closest integer.

After quantizing, the sample can be encoded, a process of physically representing the waveform information as a series of high/low electrical pulses, for example. Pulse Code Modulation (PCM), for example, samples a waveform at 8,000 times per second, and represents each measurement as an eight-bit symbol. That leads to the signaling rate for PCM voice signals of 64,000 bits per second (64 kbps). Note here that PCM, though called a "modulation" technique, actually is a coding method. The modulation is "baseband," meaning that the electrical symbols are directly impressed onto the carrying medium.

Chapter 8

The Internet

I think that we are all now convinced that the Internet is here to stay. It is not a fad, and it will remain part of the landscape for a long, long time.

This being so, it is important that we all understand the Internet, and learn how to use it to our advantage.

What is the Internet?

The Internet is simply a group of connections between computer networks. It is a large group of high-speed optical fiber lines that connect thousands of computer centers together. These communication lines are owned by AT&T, Sprint, MCI, UUNet, and several other companies, who sell access to the lines. These computer sites then connect to millions of individual homes and offices via regular or digital telephone lines.

No one really owns the internet. Every piece of the net is owned by someone, but their little piece is not necessary for the entire net to operate. There are no charges on the Internet itself: no long-distance charges, no on-line charges and no dues. The people who connect you to the net charge for *access* through their system to the internet (Compuserve, AOL, Prodigy, MindSpring, and others); and the owners of main internet lines charge for *access* to the network, but once you get in to the net, there are no further charges.

In addition to this, there are no pre-set paths for communications on the net. There are no switches to connect your computer to some other computer. There is no central brain. When you send an internet message, your computer begins the message with an electronic address. When your message reaches the next junction point, the machine there (called a router) reads the message, identifies the best path towards that destination, and sends it along. When your message reaches the next crossroad, the same thing happens. Your message may hit five or ten crossroads before it gets to its destination, and when your friend on the other end replies to your message, his or her message will probably come back to you along a different path. But as long as the addresses are valid, the messages will get to where they belong. You can see from this why the internet is called a routed network, and how the telephone system is called a switched network.

How It Began

The Internet actually began in the early 1960s, to serve the needs of researchers, and of the Pentagon, who decided that they needed a communications system that could survive a nuclear holocaust. They proposed a new concept in system structure: the center-less network. This network would be composed, not of a typical hierarchical pyramid of systems arranged in a pecking order, but an infinitely more flexible *peer-to-peer* interconnection in which no single system was in charge. Each element in the network was autonomous and independent.

That first network was called ARPANET, and at universities, it was discovered by computer students, who learned how to use it for their own benefit. They told their friends about it, and started to share notes on topics that interested them. Then, their professors discovered how to take advantage of this, and began sharing research notes with each other. Thus popular use of the internet began with academics and adolescents, which gave the net a lot of its initial flavor.

The original ARPANET was decommissioned in 1989, but by that time, there were a lot of students and professors at colleges using it, and they kept right on using it. Since 1989, the left-over internet has been growing by more than 10% per month - and continues to!

Multiple Paths

The Internet provides innumerable paths for messages to take. It is not at all unusual for a message between New York and Chicago to travel by way of Montreal, Salt Lake City, and Dallas. Many times, a message from one US city to another makes its way through a European city. That is why the net is bomb-proof; if one link closes, there are plenty of other paths to take, and of which will work just fine, and at no additional cost. And since the links are almost all composed of optical fiber, bandwidth and distance are not a problem. So, sending a message from Boston to Baltimore can go through New York City, or it can be routed through Paris, and there is no difference - no one is charged more or less, and delivery time is the same. In fact, there is no difference at all. The actual route that a message will take depends on how busy the various links are. Traffic is routed to the least busy path, no matter how long or short. If the path is less crowded and goes in the general direction of the address, that is the one to which the message will be routed.

Is the *NET* the same as the *WEB*?

Almost.

The *net* refers to the internet, the physical connection between computer networks all over the world.

The *World Wide Web* is not a separate set of connections. The "Web" is simply a special method of sending information over the net. There is no physical difference between the "net" and the "web". There is only a difference in communication protocols.

Web communications are based on the *hypertext* signal transfer system, which makes it very easy for users to interact with the network, and to share graphic files with each other. You will notice that all web addresses (also called *URLs* - Universal Resource Location) begin with "http"; which stands for *hypertext transfer protocol*.

The special hypertext communication language allows you to easily browse through files on another computer by clicking on the file-name. The hypertext system allows people to establish their own home on the web - a personal page or set of pages that people can visit, as they could visit your office or home - getting messages, giving messages, and look through any information you have posted there. You have heard these called *Web Pages*.

For these reasons, the web has become very popular.

The Importance

In a few short years, the internet has gone from hobby use to serious commerce. Almost every major company in the United States, and most of the mid-sized companies, has a web site. These sites are used in a wide variety of ways (ordering, product information, contacting people in the company, getting help with product, etc.), and are evolving in a number of directions. Here are the things that are currently being done on the net:

1. Email (electronic mail). This is the thing that really made the internet. In the lingo of the computer world, email was the internet's first "killer app" for business. *Killer app* is short for "killer application" - something you see someone else using and say "I've got to get that!" Email allows you to exchange messages with people at the convenience of all parties (no interrupting people or being interrupted), sending the same message effortlessly to any number of people, transferring documents or computer files effortlessly and instantaneously between parties, and allows you to do all of your communicating at your convenience. In real life, email messages get answered far quicker and better than phone messages or faxes. The reason is that it is a lot easier to answer an email than it is to write a letter or make phone calls. It's also a lot cheaper than regular mail (dubbed "snail mail" on the internet) or phone calls. Sending an email to Asia costs the same as sending one to the next office. If you don't have email, you are quickly becoming out of date.

2. Web sites. At the very minimum, a web site allows your customers to get almost all of their questions answered with no effort on your part.

3. Intranets. IntRAnets are built by connecting two or more offices in separate locations over the intERnet. For example, there is a company with an office in Chicago and an office in Los Angeles. In the past, if they had wanted to connect the two offices (so that people could share information quickly), they had to lease expensive telephone lines (if they could get them, and install some fairly sophisticated equipment. An alternative was to rent satellite time. Now, by connecting the two offices over the internet, they can transfer the same information without the expense of the leased line or the satellite system.

4. Encryption. This is one of the newer things on the net. Encryption refers to coded transmission - messages that are completely secure. The current generation of encryption is so secure that the fastest supercomputers in the world would have to work for years to decode a short message. There is no other form of communication that can offer this kind of privacy.

Spontaneous Generation

The sprouting of the Internet caught everyone by surprise. Not even the sci-fi writers saw it coming. A few programmers at the Dept. of Defense started it off, new communication technologies made its expansion possible, and independent student programmers in Illinois and Sweden turned it into what it is now. The Internet is a wide-open, unstructured forum. It is the Wild West of information. Governments around the world are trying to figure out what to do with it (how to regulate and tax it), and reporters around the world are trolling for Internet horror stories (proportionally, there have been an exceptionally small number of problems).

The Internet has changed the world substantially already; but this influence is by no means complete. Corporate and personal use of the net will increase. The amount of commerce done on-line is increasing rapidly, and will continue to do so. Eventually, the internet will combine with or overrun the telephone and cable TV industries.

How To Use It

First of all, the net allows you to communicate from office to customer to job site far better than you ever have before. Think about how much time and effort would be saved if you could get instant answers to all your foremen's questions on change orders, installation details, material delivery dates, and so on? (I've done the labor studies, and I can tell you that it comes out to a lot of money.) The Internet and wireless technologies allow you to take care of most of this, if you learn to adapt it to your business. Yes it takes time and effort, but there are huge benefits to be gained.

If they are not yet, your customers will soon be demanding that you have a good company web page, that your normal form of correspondence is email, and that you submit pay requests electronically. You will also find that your suppliers will give you better pricing and/or service if you order electronically. Before long, almost anything that *can* be done on-line *will* be done on-line.

Finally, there is a whole lot of data cabling being done, and there will be more. There is (and will be) a lot of specialty work. Every new type of transmission system requires specially-trained installers, requires maintenance, and requires power circuitry. There is a lot of work to be done here, but along unusual paths.

Here To Stay

The brightest, most productive, wealthiest, and most profitable people will be doing business over the net. If you can't join them there, they'll be looking for other contractors who can.

Have you been hearing non-stop about the Internet for a long time? Well, get used to it, because this is not a fad that will go away. In fact, when all is said and done, the Internet (or whatever it is called a couple of decades from now) will affect the way people live far more than you currently imagine.

The Internet's Flavor

The Internet has its own culture. This culture was especially strong in its early days (when only the true cyber-heads knew how to use it), but still exists. The atmosphere is best described as *Laissez-Faire*. This is an old French term that loosely translates to "leave us alone".

Some typical attitudes tend to be "Do whatever you want to, so long as you don't hurt anybody." "Weird opinions are OK, but so are scathing replies".

The good, the bad, and the ridiculous are all to be found on the net; it is anarchic, and the whole of human experience is found there.

Some Internet Terms

It is important that you understand the terms that we will be using. Here are some of the most important:

Bandwidth. The maximum amount of information that can be transmitted at any given time. A 56k leased line connection, for example, has 56k of bandwidth.

Client. A program that is run by users on their machine. It issues requests to a server, which is generally located on another system. This takes a big work-load off of the server program, so that it can process client requests more efficiently. This also makes the system appear very fast.

CSLIP (Compressed SLIP). SLIP with compression for a more efficient connection. See SLIP.

ECPA (Electronic Communications Privacy Act). A law passed a few years back that says that all electronic mail cannot be read by the people running the system. Its main concrete achievement seems to have been the placing of a notice on all bulletin board systems (BBSs) saying that there is no private mail function on their systems, despite the continued existence of same on the menu.

Flame. An insulting message, normally with little real content. A Flame War is a seemingly endless exchange of such messages.

FTP. File Transfer Protocol. This refers to a protocol describing the way files can be transferred over a TCP/IP network, such as the Internet. The program used to implement this protocol is also called FTP. Normally, a FTP program is included with basic networking software, and little needs to be done to make it work on a system.

HTML. (HyperText Markup Language). This is the scheme used to design World Wide Web pages. There are numerous tools that can help you write HTML with reasonable efficiency.

HTTP. (HyperText Transfer Protocol). This is the protocol used for information transmitted over the World Wide Web (WWW).

InterNIC. The government-funded service, run by a company called Network Solutions, that parcels out IP addresses and domain names.

Complaints about slow service have been heard quite loudly, and the entire system is being revised.

IRC (Internet Relay Chat). This is a direct interactive way for people to hold conversations using the computer.

Java. This is a programming language created by Sun. It was originally designed for the development of proprietary information appliances, but is now being widely used on the web.

Microsoft Internet Explorer. Microsoft's "browser" software - software that make surfing the net simple.

MOSAIC. The first truly effective web browser.

MUD or Multi-User Domain. These are the 'virtual worlds' created by serious Internet users. They were originally developed by game-players, but are now being used for commercial purposes.

Netscape. The Web Browser that made the Web a hit. Now competing with Microsoft Explorer.

News, also called **USENET**. This is a messaging system that is one of the most famous and popular parts of the net. It functions as an electronic bulletin board. Each of the thousands of newsgroups are devoted to a particular topic.

PPP. Point-to-Point Protocol. A newer and supposedly better way to connect your site to the Internet via a single serial line. Windows95 has greatly expanded its popularity, since it supports PPP instead of the older SLIP. See SLIP.

RFC. Request for Comment. This is an informal system for proposing Internet standards. The technical people who work on the Internet upload RFCs to the NIC, where they are given a number and published. Many of them are later adopted as Internet standards.

Search Engine. As the World Wide Web has grown bigger and bigger, programs have been created that wander the web, looking for resources of interest. They then put them in enormous keyword dictionaries and let you search for what you'd like to find.

SLIP. Serial In-Line Protocol. This is one of several ways to attach a computer to the Internet via a simple (and cheap) modem connection. See the earlier discussion on connecting your system to the Internet for additional information.

TCP/IP. Transmission Control Protocol/Internet Protocol. The protocol used to send information through the Internet

TELNET. Telnet is a program that lets you remotely log in to any other system on the Internet (assuming you have access).

WWW (World Wide Web). This is probably the best Internet browsing system - certainly the most fun one to use.

Encryption

As we mentioned earlier, Internet commerce could not be widely used unless messages and transactions could be assured privacy. To stop the theft and unauthorized use of data, Internet users have turned to the science of cryptography, which encrypts and decrypts information so that it is useless to everyone except authorized users.

Traditional crypto systems are based on the secret key, or symmetric, model. In this system, each user has his or her own secret key, which is used for both encryption-in which plain text is converted to ciphertext-and decryption-in which the process is reversed.

The larger the key size, the more complex the algorithm and the more difficult the crypto system is to crack. Within the United States, there is no limit in terms of key size, with many organizations going with key sizes as large as 4,096 bits. Currently, however, Federal Law prohibits the export of keys of more than 40 bits.

Because secret key crypto systems rely on a single key for both encryption and decryption, making sure all parties involved have the key, without it being intercepted, becomes a problem. Handing out the key on a disk is a relatively reliable way to go, but even that method can be uncertain.

To avoid having to place trust in a third party and to decrease the possibility of secret keys falling into the wrong hands, a new type of system, *public key cryptography,* was developed.

Public Key Cryptography

Under the public key cryptography model, each user has a pair of keys that are complementary and mathematically related. The keys are generated by a mathematical *algorithm* (formula) that generally involves very large prime numbers. With a public key system, information encrypted with a particular public key can be decrypted only by using the corresponding private key.

You can hand out your public key to each recipient you communicate with; alternatively, you may choose to post your public key at some central location or on a network directory, so anyone can look it up. (To see a working public key directory, visit Pretty Good Privacy's Web site at www.pgp.com.) By contrast, you keep your private key, and you should take care to store it in a safe, secure place, whether on a floppy disk or a secured hard drive.

When you want to securely communicate information to another party, you must first generate a key pair, if you don't already have one. Once your keys have been created, they can be used to send information through e-mail, for example. You first look up the public key of the recipient or have the recipient send the public key directly to you. You then apply the recipient's public key to the *plaintext*, or unencrypted, message and create a ciphertext, or encrypted, message, which is then sent to the intended recipient. The recipient applies his or her private key to the ciphertext to transform it back to readable plaintext.

Because Internet mail travels through numerous unknown servers to reach its final destination, it can be intercepted along the way by those with the know-how. However, unless these e-mail hijackers can provide the appropriate private key, all they will see is gobbledygook.

So, public key cryptography has several advantages over secret key cryptography. First, half of the cryptographic scheme (private keys) is never divulged, thereby increasing security. Also, public key cryptography allows for digital signatures, which enable sender and recipient to feel more confident about communicating with one other. Public key cryptography also has its drawbacks, most notably the fact that it's much slower than the secret key models. Because both types of cryptography have plenty of pros and cons, a combination of the two methods may be the most practical solution for electronic communications. You could encrypt private files, such as those on a hard drive, using the secret key method because technically those files aren't communicated to others. Then, for environments in which many people potentially have access to information, such as e-mail, you could use public key cryptography to secure information.

Remember that there are huge numbers of people on the Internet. And any time you get millions of people together, you are going to have some who are going to be harmful. Some type of cryptography, whether it's secret key or public key, will keep the wolves at bay.

Chapter 9

The Business

Understanding the technical aspects of voice, video, and data communications is certainly important, but they are only the first step when it comes to actually making money in the data communications business. Using your knowledge to make money requires a separate group of skills.

In most ways, datacom contracting is very similar to electrical contracting. Both types of contracting involve:

1. Reaching customers, and convincing them to use our services.
2. Identifying the customers' needs, furnishing them with prices, and entering into transactions.
3. Performing services according to the customer's needs.
4. Completing the terms of our contracts (collecting our money).

But communications contracting differs from traditional electrical contracting in how all of these things get done.

Getting Business

Before you can make money or employ people, you must first get jobs.

A frequent source of datacom jobs is subcontracting from electrical contractors. When an electrical contractor takes on a large project, there will usually be some fiber optic, networking, or other type of datacom work. This is included in Division 16 of the contract documents; and many electrical contractors subcontract this work to a specialist. If your company can come up with a good price, they may subcontract the work to you. This is an excellent way to get started in datacom work.

Another source is your existing customers. Although the amount of work you can get from your customers will vary, almost all of your commercial and industrial customers use communications systems, as do many residential customers. These customers typically upgrade their systems every several years.

You may also get referrals from distributors and component dealers. These people often have customers asking them about installers. If your company supplies the vendors with some sales and promotional information, you may get some very good referrals.

Advertising

One of the primary methods of advertising is the company brochure. This is perhaps the most general form of advertising; in that it can be given to any potential customer.

Another common type of advertising is direct mail. This entails communications contractors sending promotional literature to potential customers in their area. Then, the mailing may be followed with telephone calls to the recipients, and then with another mailing. (Perhaps with more phone calls also.) Add campaigns such as this take time. Results are often slower than you might hope, but are usually fairly successful.

Like maintaining a trained workforce, advertising is an expense that will make you money. For this reason, mark-up levels must be increased, and additional overhead time and expense will be incurred.

Training

Getting into datacom work requires you to think hard about obtaining properly trained employees. If you wish to use your existing employees, you will have to train them in the systems they will be installing. To do this, you will have to enroll them in some type of training program, whether it be an intensive seminar-type course, a community college program, or correspondence training.

Your other choice is to hire people who are already trained. Eventually, there will be a sizeable pool of fully-trained communications installers; but until then, the contractor must grab the best ones you can when they hit the marketplace. This makes the job and expense of maintaining a workforce quite a bit higher than it is for power wiring. This is reflected in the higher mark-ups that are usually charged for communications work. It may also require extra overhead employees.

There are several other options available to you for training installers:

Local community colleges provide excellent training, but take a long time, and require a serious time comitment (after work). This usually disqualifies them from consideration if installers must be trained quickly.

Traveling training programs can also be excellent, but are only available at certain places and times. They can be expensive, but they will provide your new installer with a pretty good knowledge base in only a few days.

Correspondence courses. There are a few available. Correspondence training is a nice adjunct to on-the-job training, since it allows the new technician to study at night, and discuss the lessons with his or her trainer. It is also relatively inexpensive.

If you are a union contractor, you may (or may not) have acces to trained installers. If so, you will have to do relatively little on-the-job training. Also, you may be a non-union contractor with access to a good apprenticeship program, and can get well-trained people that way. In either of these cases, you will have to train new people in your company's methods of test documentation, wire marking, and perhaps a few other specifics.

Keeping employees is also an issue in the datacom business. Your best workers may wish to move up to better jobs. In the datacom field, that generally means that they wish to become network managers for a large company, or to work in a start-up firm. There is no good way to keep these people in your employ, but you can make the situation better by offering to help any of your workers that wishes to move up. Ask them about their future plans, and help them reach their goals. Simply ask that in return, they keep you informed of what they want to do, and help train their replacements.

Estimating

This section will cover the basic techniques of datacom estimating. Many of these steps are common with electrical estimating, but they bear repeating.

The first step in estimating is to ascertain the overall requirements of the job being estimated. You must get a clear picture in your mind of how this job will flow; and more importantly, where will the money come from, and when. In addition, you must understand the scope of the work for which you are quoting a price, and exactly what will be required of you. These are primary concerns, and are the first considerations in any good estimate.

To verify that all of these factors are considered, many estimators use checklists that they review for every project. You should develop your own for such uses.

Special attention must be given to the supply channels through which the materials will flow. Conventional electrical estimating requires attention to be paid to this matter, but not nearly to the same extent that is required for communications work. The reason for this is simply that the distribution system for electrical construction materials is highly developed and compact. Finding the materials you need, when you need them, is not generally too great of a problem.

The flow of materials for electrical power installations flows as follows:
1. The manufacturer buys raw materials form whatever sources are appropriate, processes, assembles and fabricates them until the product is complete.
2. The materials are placed in a warehouse until they are shipped to whatever electrical wholesalers order them.
3. To stay in contact with the hundreds of distributors scattered throughout the country, nearly all manufacturers use a network of manufacturer's rep firms. Each of these representatives works in a certain geographical area, promoting the manufacturer's products and taking orders. The rep firms (called agencies) are paid a percentage of the manufacturer's product sales in their area.
4. The electrical distributors (local supply houses) send out sales people to call on individual customers in their area. These sales people are paid on a percentage of their sales, and therefore have a big incentive to get the customer whatever the customer wants.

While there are number of steps in the process we have just gone through, you can see that it is a very efficient process. The end buyer (the electrical contractor) has a very convenient channel of obtaining the materials he needs. And, should there be a difficulty in obtaining certain materials, the running around and phone calls are done by the wholesale house's salesman. Thus the contractor can usually deal with the ordering process just once, with only one salesman.

The datacom purchasing process is not always so simple. One or more links in the distribution chain may be missing.

Because of these gaps in the distribution chain, the processes of estimating and purchasing for communications work are made more difficult. More user-friendly product distribution methods are being developed; but extra difficulty is still frequently involved.

The distribution channels for datacom materials are varied and scattered. This impacts directly on the estimator. To make the problem somewhat easier to handle, you can use estimating forms that have columns for listing vendors.

Datacom estimating requires you to evaluate your sources of supply during the estimating process. Where the material will be coming from is a serious concern. It is a risk you must attend to, if you want reliable estimates.

The Take-Off

The process of taking off communications systems is essentially the same as the process used for conventional estimating. By taking off, we mean the process of taking information off of a set of plans and/or specifications, and transferring it to estimate sheets. This requires the interpretation of graphic symbols on the plans, and transferring them into words and numbers that can be processed.

Briefly, the rules that apply to the take-off process are as follow:

Review the symbol list. This is especially important for communications work. Communications systems are not standardized; and therefore vary widely. Make sure you know what the symbols you are looking at represent. This is fundamental.

Review the specifications. Obviously it is necessary to read a project's specifications, but it is also important to review the specifications before you begin your takeoff. Doing this may alert you to small details on the plans that you might otherwise overlook.

Mark all items that have been counted. Again, this is obvious, but a lot of people do this rather poorly. The object is to clearly and distinctly mark every item that has been counted. This must be done is such a way that you can instantly ascertain what has been counted. This means that you should color every counted item completely. Do not just put a check mark next to something you counted; color it in so fully that there will never be any room for question.

Always take off the most expensive items first. By taking the most expensive items off first, you are assuring that you will have numerous additional looks through the plans before you are done with them. Very often, you will find stray items that you missed on your first run through. You want as many chances as possible to find all of the costly items This way, if you do make a mistake, it will be a less expensive one.

Obtain quantities from other quantities whenever possible. For example, when you take off conduit, you don't try to count every strap that will be needed. Instead, you simply calculate how many feet of pipe will be required and then include one strap for every 7-10 feet of pipe. We call this *obtaining a quantity from a quantity.* Do it whenever you can and it will save you good deal of time.

Do not rush. Cost estimating, by its very nature, is a slow, difficult process. To do a good estimate, you must do a careful, efficient take-off. Don't waste any time; but definitely don't go so fast that you miss things.

Maintain a good atmosphere. When performing estimates, it is very important to remain free of interruptions, and to work in a good environment. Spending hours counting funny symbols on large, crowded sheets of paper is not particularly easy; make it as easy on yourself as you can.

Develop mental pictures of the project. As you take off a project, picture yourself in the rooms, looking at the items you are taking off. Picture the item you are taking off in its place, its surroundings, the things around it, and how it connects to other items. If you get in the habit of doing this, you will greatly increase your skill.

Labor Expense

The following list contains the essential elements of jobsite labor. Any good cost estimate must cover these operations. The conventional method of estimating does this conveniently by assigning a single labor unit to each item.

1. Reading the plans.
2. Ordering materials.
3. Receiving and storing materials.
4. Moving the materials to where they are needed.
5. Getting the proper tools.
6. Making measurements, laying out the work.
7. Installing the materials.
8. Testing.
9. Cleaning up.
10. Lost time and breaks.

The conventional method of estimating is excellent where circumstances are consistent and predictable. But with communications estimating, there are other factors that come into play. For instance, we can specify a certain labor rate for controller, but there may be other functions required along with the controller, such as programming, software installation, testing, modifications to programs, etc.

Because of this situation (which is typical to most types of communications equipment), a single labor unit is not sufficient to assign labor for the item. What might be applicable for one installation may be double what is appropriate in the next installation - even though it is exactly the same type of equipment. The challenge of communications estimating is finding a method of assigning labor in either situation.

To solve the problems of charging communications labor, several solutions have been put forward. While all can work well, I recommend that you simply use two lines on the estimate form for each item - one for a base labor unit (and the material cost), and the second for "connect" labor.

As you saw in the list of the elements of labor, there are ten basic functions that we bundle together into one labor unit. Our solution to the high tech labor problem is to isolate and remove the variable element from the other nine elements of labor, and charge for it separately.

By removing operation number 8 (testing) from our labor unit, and charging for it separately, we clean up the estimating process. All of our volatility is moved into one separate category, which demands thought before a labor cost is applied; and the rest of the labor, which is consistent and predictable, is covered by the basic labor unit.

As may be obvious, the title "Testing" is hardly an accurate term to describe the types of operations our second labor figure is to account for. Testing would be only one component of these operations. Along with testing would be included any number of technical operations, such as the following;
Installing software.
Programming.
Configuring hardware.
Configuring software.
Training users.
Creating reports.

Because of the many operations that could fit into this category, there seems to be no good name to give this category. So, we are simply using the word "connect labor" to identify it.

"Connect" labor is where all of the volatility of communications estimating goes. Connect labor should be calculated and charged separately from normal labor, which we are calling "installation labor". Installation labor is the labor required to procure the material, move it to the installation area, tool up, and mount the equipment in its place. In short , all of the normal elements of labor, except for the "Connect" labor.

Modifying Labor Units

Labor units are based upon the following conditions:
1. An average worker.
2. A maximum working height of twelve feet.
3. A normal availability of workers.
4. A reasonably accessible work area.
5. Proper tools and equipment.
6. A building not exceeding three stories.
7. Normal weather conditions.

Any set of labor units must be tempered to the project to which they are applied. They are a starting point, not the final word.

Unusually poor working conditions typically require and increase of 20-30%. Some very difficult installations may require even more. Especially good working conditions, or especially good workers may allow discounts to the labor units of 10-20%, and possibly more in some circumstances.

Subcontracts

Datacom installations frequently require the use of specialty subcontractors. The reason for this is that there are so many new specialties, no one installer can be knowledgeable in all of them. Therefore, it is likely that all but the largest firms will encounter many situations where they have no one in their company who is knowledgeable about a certain type of application. In these circumstances, you will normally have to call the manufacturer of the system, who can refer you to a consultant or dealer of theirs in your local area, who can do this work for you.

Charging Training

Many datacom installations require you to teach the owners or their representatives to use the system. This is worse for high tech work than for power wiring. After all, you don't have to spend time teaching building owners how to use conventional electrical items, such as light switches, but you will certainly have to spend time teaching them how to use a sound system. Not only that, but you may have to supply operating instructions, teach a number of different people, and answer numerous questions over the phone after the project is long over.

The question here is whether you include training charges in "Connect" labor, or include these cost as a separate job expense. This decision is essentially up to the discretion of the estimator. It is, however, usually best to charge general training to job expenses, and to include incidental training in "Connect" labor.

If you come from a background in the electrical construction industry, make sure that you accept these expenses as integral part of your projects. Do not avoid them. The people who are buying your systems need training, and they have a right to expect it. Include these costs in your estimates, and choose your most patient workers to do the training.

Overhead Percentages

What percentage of overhead to assign to any type of work can be a hotly debated subject. Everyone seems to have their own opinion. Whatever percentage of overhead you charge, consider raising it a bit for communications projects. As we have already said, the purchasing process is far more difficult for communications work than it is for other, more established types of work. In addition to this, there are a number of other factors that are more difficult for communications work than they are for more traditional types of work.

Almost every factor we can identify argues for including more overhead charges for communications work. Not necessarily a lot more, but certainly something more. When you contract to do communications installations, you are agreeing to go through uncharted, or at least partially uncharted, waters. This involves greater risk. And if you do encounter additional risk, it is only sensible to make sure you cover the associated risks. You do this by charging a little more for overhead and/or profit.

Bidding

There are many types of bids that you will be asked to furnish in the communications market. Most of them will be lump-sum bids, or unit-priced bids. But there is another type of bid that you will encounter, the *RFP*.

RFP stands for *Request For Proposal*. In many specialties (and in the data networking business in particular), RFP is almost equivalent to bid documents, except not as detailed. When these people want a bid from you, they will send you an RFP. This document will give you the general details of the project, and ask you to furnish a complete design, schedule, and price.

Completing the RFP process is very similar to performing a design/build proposal.

When you prepare RFPs, take care that you will not be doing the design work for the customer (for free), only to have another party do the installation according to your design.

A New Company

When deciding to do datacom work, you must first verify (as best you can) the viability of your choice. In all these matters, I think you'll find that data communications fits the criteria - but you should verify the following before jumping in:

1. Amount of work available, current and future. You want to choose a specialty that is growing, and will continue to grow for a long time. As long as growth is occurring, supply and demand should remain in your favor.

You can determine this by doing a bit of research, and finding projections for the various trades. To get the projection numbers you want, you can:
- call or visit a reference librarian, ask the librarian to perform a search of their various databases,
- find trade journals in that specialty, and search for answers,
- call trade organizations in that specialty, or
- post requests for information on special computer bulletin boards.

You want to find businesses that are already busy, and will continue to grow. Do not choose a business that is almost nonexistent now, but will be huge later - you need something now.

2. Skills and training. Step number two is to find out how much difficulty you will have in training yourself and your workers to perform this work. Your first level of training is for yourself, so that you will be at least as proficient as any of your employees. Your second level of training is for your employees.

Calculate your time and expense to get everyone properly trained.

3. Equipment and tools. Identify the types of equipment and tools you will need to get started in the new trade. Specify prices for each item, and come up with a total; it doesn't have to be as accurate as a bid, but it should be reasonably close to reality.

Remember that as you start-up, you may be able to rent expensive pieces of equipment and tools, rather than purchasing them. For instance, if you wish to get started in data cabling, you don't need to buy all of the expensive testers: Just verify that they are locally available on a rental basis, and eliminate them from your "must purchase" list.

4. Determine supply channels. Next, you must identify your supply channels. We've been spoiled with electrical materials - there is no difficulty at all in finding almost any electrical item we might desire. Not so in the high tech specialties. It is not terribly difficult to find what you want in these trades, but it is certainly not as easy as running by the supply house on the way to the job.

Make a list of the suppliers that you will need to do business with to perform your specialty work. Pay attention to delivery times. For instance, if you have to buy mail-order, you will have to order materials long in advance of the installation time. On the other hand, items that can be bought from a local supplier can be ordered just a day or two in advance.

5. Make a list of local customers. Your next step is to prepare a list of customers in your area. Invest some effort in this, and come up with a good list.

6. Define your marketing techniques. How can you best reach the people on your customer list? Consider options such as direct mail, sales calls, bidding, and so on. Also try to determine what your costs are likely to be for a serious marketing campaign.

7. Analyzing the data. Now, you will have to analyze the data you've developed. Rate each specialty you are considering in each of these six factors. At the end of your analysis, you should be able to come to an intelligent decision as to which specialty or specialties you will enter.

Company Structure

Once you have made the first essential decisions, you need to decide how your new company will be structured.

It is certainly possible to run the new types of work within your existing company, but as a practical matter, it is almost always better to separate the two different types of work. Even if you will still be using the same shop, installers, and tools, you will certainly want to keep separate sets of books. Those contractors who are already doing this have almost all gone to a separate company structure.

For general business and financial reasons, the new company should be incorporated, not a proprietorship or a partnership. (Limited partnerships sometimes work, but require serious legal expenses just to set up.)

Now the hard questions: Who will be the owner of the company? And if more than one person, who owns how much? How much stock will be offered? These questions can be difficult. Get a lawyer to handle them for you. The lawyer's advice in these matters is necessary.

Infrastructure

Setting up a new company can be a lot of fun; it allows you to build the perfect company from the ground up. You can do in this company everything you've always wanted, with no resistance from the status quo.

What you want to do is to set up a system of operating that covers everything that the company needs done. Make a list of all the activities that will be necessary (marketing and sales, estimating, billing, collecting, accounting, purchasing, answering the phones, supervision, etc.). Then assign each of these activities to a specific person. Then compare the lists, see how it will all work together, and modify where necessary.

If you can assign all of these duties, and provide each employee with a list of responsibilities, your company will be easy to run.

Also identify the computer programs you will use, and what types of computers you will need. Try to get as much of your work as possible off of paper and on to disk. If you do this from the very beginning, it shouldn't be especially difficult.

At this time, you will have to make arrangements for your location. Most contractors like to run their spin-off company from their existing shop, to avoid new rental and utility expenses. (When starting a new company, it is generally advisable to avoid spending any extra money.) If, however, you have no more room, you will have to rent a separate facility. This will, however, increase your start-up costs significantly.

You must also set up your estimating and job pricing methods. Estimating these projects is different than power work. Be sure that you know what you are doing before you start throwing prices around. No matter how good your market may be, if you quote bad prices, you'll get stung.

How Long It Takes

This is a lot of work. Starting any type of business should be taken very seriously. The goal of this research is to eliminate all possible risks, and to define the best methods of conducting business, before actually spending any money.

The first part of your research and planning shouldn't take you more than a day or two; although gathering your market projections may include many days of waiting. The legal matters may take a few weeks before they are completed by the attorney.

Once you have all of your planning done, you can prepare to conduct commerce. Your first duty is to make sure that you are both able and available to perform the actual installations. Get the necessary tools and/or equipment, and make sure that your people are trained and ready to go.

At this point, you will have to have your shop (or shop area) ready to go as well. Have the materials and tools laid out in a sensible manner.

Once you are ready and able to conduct business, you can begin to market your services. I strongly suggest that you do not begin marketing until your ability to perform the services is assured; one of the worst things you can do in new markets is to promise something you cannot deliver.

In most cases, this will mean that your people will be all ready to go, and will then have to wait for a job to roll in. Tell them in advance that this will happen, and it should cause no hardships.

First Steps

Prepare yourself for a slow start. New companies take a while to get moving. If your first adds don't secure any business, try a different angle, and keep going. Don't get scared if your first results are slow in coming. If you've done your research well, your work will soon be rewarded.

Once you begin to get work, be sure to go very slowly. These are new types of projects to both you and your workers. There are certain to be many new, confusing, and difficult situation that will arise. You will need time deal with these as they appear. If you move slowly, you can resolve these difficulties without creating problems for your customers. If, however, you go too fast, you will not be able react to problems fast enough, and you will disappoint your customers.

Once you get past your first series of projects, you can begin to expand. Even so, do not expand too quickly, as handling a number of projects at the same time will also present obstacles to be overcome. Don't be in a hurry, and in a year or two you can be very profitable.

Supervising

When you begin to take on projects, it is critical that you know how to supervise the work your people do. Since we are familiar with power wiring, we think of supervision primarily as making sure that the work gets done on time. Making sure the work is done right is a consideration as well, but it is not the biggest thing in our minds. With communications work, however, getting it right is more difficult than getting it done in time.

The problem is this: Mistakes in communications work (terminations in particular) are difficult to detect. A mistake that could keep the entire system from working might not show up at all until the system is completely installed and turned on. (Think about that for a minute - it's a scary situation.)

So, properly supervising a communications installation means that you must be able to make sure that your work is good. Yes, it must be done on time; but on time is meaningless if it has to be replaced.

Job number one is to assure that all the terminations are done correctly. This is by far the most important part of supervision.

Make sure that your people have all the right parts;
make sure that they don't rush;
make sure that they have a well-lit work area;
make sure that they have test equipment;
and make sure that they use it!

Next, be very sure that they mark every run of cable and termination well. Waste money on cable markers and numbers;
waste time on written cable and termination schedules;
but DO NOT lose track of which cable is which.

Look over the shoulders of your people on the job to make sure they are doing things right. Terminating communications is fine work, make sure that your people work like jewelers, not like framing carpenters.

Inspection

Inspection is a bit of a wild card in datacom installations. Electrical inspectors don't always inspect communication wiring. Nonetheless, take a moment to check with a local electrical inspector before you do work in their jurisdiction. And obviously you should be very familiar with the requirements of Article 800 of the NEC.

In most cases, the inspector of your communications installation will be the same person who signs your contract. In some cases, it will be a third party.

No matter what, make sure that you know who will inspect your work before you give your customer a final price. You must know what the inspector will expect of you, and what he or she will be looking for. Be especially careful of third party inspectors, since they are getting paid to find your mistakes.

The bottom line in installation quality is getting good signal strength and quality from one end of the network to the other. But be careful of other details that may be noticed by the inspector. Among other things, many inspectors will give a lot of attention to proper cable marking, mechanical protection, and workmanship. Pay attention to any detail that the inspector is likely to look for.

Business Skills

Although this book does not deal with general business skills, they are critically important to you. You need to understand finances, and where wealth comes from. Here is a brief list of books you should read to educate yourself in money and wealth matters:

THE RICHEST MAN IN BABYLON, Clason
ATLAS SHRUGGED, Rand
THINK AND GROW RICH, Hill
MILLION DOLLAR IDEAS, Ringer
THE 22 IMMUTABLE LAWS OF MARKETING, Reis & Trout

Glossary of Terms

- A -

A & B Leads: Designation of leads derived from the midpoints of the two pairs comprising a 4-wire circuit.

A/D Converter: Analog to digital converter.

Abbreviated Dialing (AD): Preprogramming of a caller's phone system or long distance company's switch to recognize a 2- to 4-digit number as an abbreviation for a frequently dialed phone number, and automatically dial the whole number. Synonym: Speed Dialing.

Access Charge (AC): Monies collected by local phone companies for use of their circuits to originate and terminate long distance calls. Can be per minute fees levied on long distance companies. SUBSCRIBER LINE CHARGES Charges levied directly on regular local lines, fixed monthly fees for special telephone circuits.

Access Line (AL): A telephone circuit which connects a customer location to a network switching center.

Accessible: Can be removed or exposed without damaging the building structure. Not permanently closed in by the structure or finish of the building.

Accessible, Readily (Readily Accessible): Can be reached quickly, without climbing over obstacles or using ladders.

Address (Computer): The designation of a particular word location in a data register.

Aggregate: A masonry substance that is poured into place, and then hardens, as in concrete.

All Trunks (ATB): A single tone interrupted at a 120 ipm (impulses per minute) rate to indicate all lines or trunks in a routing group are busy. The fast "all circuits busy" signal.

Alternate Route (AR): A secondary communications path used to reach a destination if the primary path is unavailable.

Alternate Use (AU): The ability to switch communications facilities from one type to another, i.e., voice to data, etc.

Alternate Voice Data (AVD): A single transmission facility which can be used for either voice or data.

Alternating Current (AC): Electrical current which reverses direction repeatedly and rapidly. The change in current is due to a change in voltage which occurs at the same frequency.

Ambient Temperature: The temperature of the surroundings.

Ampacity: The amount of current (measured in amperes) that a conductor can carry without overheating.

Ampere (Or AMP): Unit of current measurement. The amount of current that will flow through a one ohm resistor when one volt is applied.

Ampere-Hour: The quantity of electricity equal to the flow of a current of one ampere for one hour.

Analog: The representation of one quantity by means of another quantity proportional to the first.

Analog Signal (AS): A signal in the form of a continuous varying quantity.

Angle of Incidence: The angle that a light ray striking a surface makes with a line perpendicular to the reflecting surface.

Annunciator: An audible intercept device that states the condition or restrictions associated with circuits or procedures.

Approved: Acceptable to the authority that has jurisdiction.

Architecture (Computer): The conceptual design of computer hardware.

ASCII: American Standard Code for Information Interchange. A standard data code.

Attenuation: A general term used to denote the decrease in power between that transmitted and that received due to loss through equipment, lines, or other transmission devices. It is usually expressed as a ratio in dB (decibels).

Automatic Self-acting: Operating by its own mechanism, based on a non-personal stimulus.

- B -

Bandwidth: The range of frequencies between two defined limits.

Bare Wire: An electrical conductor having no covering or insulation whatsoever.

Baseband: The total frequency band occupied by all the signals used to modulate a radio carrier.

BASIC: Beginner's All-purpose Instruction Code. A computer compiler language.

Baud: Signaling speed expressed as level changes per second. 100 pulses per second equals 200 baud.

Bit: A binary digit.

Bonding: The permanent joining of metal parts to form an electrically conductive path.

Bonding Jumper: A conductor used to assure the required electrical connection between metal parts of an electrical system.

Boot, or Bootstrap (Computer): A short series of instruction codes that program a computer to read other codes. The first step in the computer's operations.

Branch Circuit: Conductors between the last overcurrent device and the outlets.

Branch Circuit, Individual: A branch circuit that supplies only one piece of equipment.

Break: A means of interrupting transmission, a momentary interruption of a circuit.

Bridle Ring: An inexpensive support device usually used to loosely hold telephone wiring where appearance is not a factor. The bridle ring screws into the supporting surface. It is usually used where the wire is run below six feet and contains no sharp or hazardous edges. The telephone wire is inserted after the ring is in place.

Broadband: A transmission facility having a bandwidth of greater than 20 kHz.

Building: A structure that is either standing alone, or cut off from other structures by fire walls.

Bus: A conductor, or group of conductors, to which several units of the same type of equipment may be connected.

Bus Bar: The heavy copper or aluminum bar used to carry currents in switchboards.

Byte: A binary digit, made up of eight bits.

- C -

Capacitor: An electrical device which causes the current in a circuit to lead the voltage, the opposite effect of induction.

Carrier: A long distance company which uses primarily its own transmission facilities. Sometimes confused with reference to resellers which lease or buy most or all transmission facilities from carriers.

Central Office: The nearby building containing the local telco switch which provides local telephone service. Also the physical point where calls enter the long distance network. Sometimes referred to as Class 5 office, end office, or Local Dial Office.

Centrex PBX: Service provided by a switch located at the telephone company central office.

Cleats: Porcelain fasteners that are used to fasten electric power wires that are insulated but do not have an outer protective jacket.

CMOS: Complimentary Metal Oxide Semiconductor.

Coaxial Cable: A cable consisting of two conductors that are concentric with each other and insulated from each other.

CODEC Coder-Decoder: Used to convert analog signals to digital form for transmission over a digital median and back again to the original analog form.

Common Carrier: A government-regulated private company that provides the general public with telecommunications services and facilities.

Conditioning Equipment: Equipment modifications or adjustments necessary to match transmission levels and impedances and which equalize transmission and delay to bring circuit losses, levels, and distortion within established standards.

Common Mode: Placed upon both sides of an amplifier at the same time.

Concealed: Made inaccessible by the structure or finish of the building.

Connecating Block: A device used for terminating premises telephone wiring and a means of connecting telephone sets to such wiring.

Control Unit: The central processor of a telephone switching device.

Core Storage: Binary memory storage, made up of tiny magnetic elements.

CPU (Computer): Central Processing Unit.

Cross Connection: The wire connections running between terminals on the two sides of a distribution frame, or between binding posts in a terminal.

Cross-Sectional Area: The area (in square inches or circular mils) that would be exposed by cutting a cross-section of the material.

Cross Talk: The unwanted energy (speech or tone) transferred from one circuit to another circuit via electromagnetic induction.

Customer Premise Equipment: Telephone equipment, usually including wiring located within the customer's part of a building.

Customer-Provided Inside Wire: Wiring which is inside the premises from a Network Interface (NI) to the telephone hardware and is provided by the customer.

- D -

Decibel (DB): A unit measurement represented as a ratio of two voltages, currents or powers and is used to measure transmission loss or gain.

Dedicated Access Line (DAL): An analog special access line going from a caller's own equipment directly to a long distance company's switch or POP. Usually provided by a local telephone company. The line may go through the local telco Central Office, but the local telco does not switch calls on this line.

Demodulation (MOD): The process of retrieving data from a modulated signal.

Digital: A system in which characters and codes are used to represent numbers in discrete steps.

Diode: Electronic component that allows current flow in one direction only. See Blocking Diode, Bypass Diode.

Direct Current (DC): Electrical current that flows in one direction only.

Disk (Computer): A rotating disk covered with magnetic material, used for storage of data.

Drive Ring: An inexpensive device used to loosely hold telephone wiring in place where appearance is not a factor. The nail in a drive ring is driven into the supporting surface and the ring is open to permit placing or wires. A drive ring must be at least six (6) feet from the floor so that its nail will not present a hazard.

Drop: The portion of outside telephone plant or power system that extends from the distribution system to the customer's premises.

Drop Wire: Wire used to transmit telephone service into a customer's premises. It may be aerial or buried.

Dual Tone Multi-Frequency (DTMF): Also known as TOUCH-TONE. A type of signaling which emits two distinct frequencies for each indicated digit.

Duplex: Simultaneous two-way independent transmission.

Duplex Signaling: A long-range bidirectional signaling method using paths derived from transmission cable pairs. It is based on a balanced and symmetrical circuit that is identical at both ends.

- E -

EDP: Electronic data processing.

EPROM: Erasable Programmable Read Only Memory.

Exchange: A telephone switching center.

- F -

FARAD: The unit of measurement of capacitance.

Federal Communications Commissio (FCC): The government agency established by the Communications Act of 1934 which regulates the interstate communications industry.

File (Computer): A block of data.

Filter: A combination of circuit elements that is specifically designed to pass certain frequencies and resist all others.

Four Wire Circuits: Circuits that use two separate one-way transmission paths of two wires each, as opposed to regular local lines that usually only have two wires to carry conversations in both directions. One set of wires carries conversation in one direction, the other in the opposite direction.

Frequency: The number of complete cycles per unit of time.

Frequency Division Multiplexing (FDM): The division of an available frequency range (bandwidth) into various subdivisions, each having enough bandwidth to carry one voice or data channel.

Full Duplex: A circuit that allows transmission of a message in both directions at the same time. Synonym: 4-wire.

- G -

Ground: An electrical connection (on purpose or accidental) between an item of equipment and the earth.

- H -

Half Duplex: A circuit for transmitting or receiving signals in one direction at a time.

Hardware (Computer): The physical computer and related machines.

Hardwire: To wire or cable directly between units of equipment.

Harmonic Distortion: The ratio, expressed in decibels, of the power at the fundamental frequency, to the power of a harmonic of that fundamental.

Hertz (HZ): International standard unit of frequency. Replaces, and is identical to, the order unit "Cycles-per-second. "

- I -

Impedance: The effects placed upon an alternating current circuit by induction, capacitance, and resistance. Total resistance in an AC circuit.

Inductance: The characteristic of a circuit that determines how much voltage will be induced into it by a change in current of another circuit.

Interface: The junction or point of interconnection between two systems or equipment having different characteristics.

Interference: Any unwanted noise or crosstalk on a communications circuit that acts to reduce the intelligibility of the desired signal or speech.

I/O, or Input/Ouput, Interface: A device that regulates data flow into and out of a computer or processor.

Intergrated Circuit, or IC: A circuit in which devices such as transistors, capacitors, and resistors are made from a single piece of material and connected to form a circuit.

- L -

Leased Lines: A circuit or combination of circuits designated to be at the exclusive disposal of a given subscriber.

Local Area Network (LAN): Intraoffice communication system usually used to provide data transmission in addition to voice transmission.

Local Loop: The local connection between the end user and the central office.

Location, Damp (Damp Location): Partially protected locations, such as under canopies, roofed open porches, etc. Also, interior locations that are subject only to moderate degrees of moisture, such as basements, barns, etc.

Location, Dry (Dry Locations): Areas that are not normally subject to water or dampness.

Location, Wet (Wet Location): Locations underground, in concrete slabs, where saturation occurs, or where floors are used as plenums.

- M -

Machine Language: Programs or data that are in a form that is immediately useable by the computer, usually binary.

Main Distribution Frame (MDF): The point where outside plant cables terminate and from which they cross-connect to terminal or central office line equipment.

Main PBX: A PBX directly connected to a tandem switch via an access trunk group.

Microprocessor: A central processing unit in the form of a large integrated circuit. A microprocessor is the central component of a computer.

Microwave: Radio transmission using very short wavelengths, corresponding to a frequency of 1,000 megahertz or greater.

MODEM (Modulator-demodulator): A device that connects a computer to a telephone line and sends data over those phone lines, normally using a modulated audio tone.

Modular: Usually applies to the connection of a telephone set mounting cord to a telecommunications network via plugs located on the end of such cords and jacks used to terminate premises inside wire.

Modulation: Alterations in the characteristics of carrier waves. Usually impressed on the amplitude and/or the frequency.

Motherboard: A board containing a number of printed circuit sockets and serving as a backing.

Multiplexing: The act of combining a number of individual message circuits for transmission over a common path. Two methods are used for electrical signals: (1) frequency division, and (2) time division. Wavelength division multiplexing is used for optical signals.

NEC: National Electrical Code, that contains safety guidelines for all types of electrical installations.

Network Interface: The physical and electrical boundary between customer premises inside wire and telecommunications network. The Network Interface can be any telephone company provided modular jack.

- O -

OHM: The unit of measurement of electrical resistance. The amount of resistance that limits current flow to one ampere, under an electrical pressure of one volt.

Overbuild: Adding capacity to a telecommunications network, usually by installing new cabling next to the old.

Overcurrent: Too much current.

- P -

Plenum: A chamber that forms part of a building's air distribution system, to which connect one or more ducts.

Point of Presence (POP): A physical location within a local area, at which a carrier establishes itself for the purpose of obtaining access, and to which the Regional Bell Operating Company (RBOC) provides access services.

Point-to-Point: A communications circuit between two terminations that does not connect with a public telephone system.

Port: Entrance or access point to a computer, multiplexor device or network where signals may be supplied, extracted or observed.

Private Branch Exchange (PBX): A private phone system (switch) used by medium and large companies that is connected to the public telephone network (local telco) and performs a variety of in-house routing and switching. User usually dial "9" to get outside system to the local lines.

Private Line: A full-time leased line directly connecting two points, used only by purchaser. The most common form is a tie line connecting two pieces of a user's own phone equipment - flat rate billing, not usage sensitive.

Protector: A device used as protection from hazardous voltages. It may be mounted either inside or outside the premises. If mounted outside it will be covered with a plastic or metal housing.

- R -

Raceway: A metal or plastic channel used for loosely holding electrical and telephone wires in buildings. A raceway may be a conduit, rectangular wireway, or any other type of protected channel.

Radius (Radii, Plural): The distance from the center of a circle to its outer edge.

RBOC {Regional Bell Operating Company): The seven resulting companies from the breakup of AT&T.

Redundancy: Duplicate equipment that is provided to minimize the effect of failures or equipment breakdowns.

Regeneration: The process of receiving distorted signal pulses and from them recreating new pulses at the correct repetition rate, pulse amplitude, and pulse width.

Repeater: A device that receives weak signals, and transmits corresponding stronger signals.

Reseller: A long distance company that purchases large amounts of transmission capacity or calls from other carriers and resells it to smaller users.

Resistance: The opposition to the flow of current in an electrical circuit.

Resonance: A condition in an electrical circuit, where the frequency of an externally-applied force equals the natural tendency of the circuit.

Restoration: The re-establishment of service by rerouting, substitution of component parts, resplicing, or by any other means.

- S -

Satellite Relay: An active or passive repeater in geosynchronous orbit around the Earth, which amplifies the signal it receives before transmitting it back to earth.

Semiconductor: A material that has electrical characteristics somewhere in between those of conductors and insulators.

Service: Equipment and conductors that bring electricity from the supply system to the wiring system of the building being served.

Service Drop: Overhead power conductors from the last pole to the building being served.

Signal Circuit: An electrical circuit that supplies energy to one or more appliances yielding a recognizable signal.

Signal To Noise Ratio: Ratio of the signal power to the noise power in a specified bandwidth, usually expressed in db.

Software (Computer): Programming, especially problem-oriented.

Station: Any location on a network capable of sending or receiving messages or calls.

Switch: Equipment used to interconnect lines.

- T -

T-1: 24 voice channels digitized at 64,000 bps, combined into a single 1.544 Mbps digital stream (8,000 bps signaling), and carried over two pairs of regular copper telephone wires. Used for dedicated local access to long distance facilities, long-haul private lines, and for regular local service.

T-Carrier: A time-division, pulse-code modulation, voice carrier used on exchange cable to provide short-haul trunks.

TELCO: Local telephone company.

Telecommunications: The transmission of voice and/or data through a medium by means of electrical impulses and includes all aspects of transmitting information.

Telecommunications Network: The public switched telephone network.

Terminal: A point at which information can enter or leave a communications network.

Terminal Equipment: Devices, apparatus and their associated interfaces used to forward information to a local customer or distant terminal.

Thermal Protection: Refers to an electrical device that has inherent protection from overheating. Typically in the form of a bimetal strip that bends when heated to a certain point. When the bimetal strip is used as a part of appliance's circuitry, the circuit will open when the bimetal bends, breaking the circuit.

Time Division Multiplexing (TDM): Equipment that enables the transmitting of a number of signals over a single common path by transmitting them sequentially at different instants of time.

Tip: The side of a two wire telephone circuit that is connected to the positive side of a battery at a Telephone Company Central Office.

Transducer: A device that converts energy from one form to another; especially a device that converts a physical quantity to an electrical current.

Transmission: The electrical transfer of a signal, message or other form of data from one location to another without unacceptable loss of information content due to attenuation, distortion, or noise.

Transmission Level: The level of power of a signal, normally 1,000 Hz, that should be measured at a particular reference point.

Trunk: A telephone circuit or path between two switches, at least one of which is usually a telephone company Central Office or switching center. Regular local CO circuits are called PBX trunks, because there is a switch at both ends of the circuit.

- U -

Uniform Service Order Code (USOC): The information in coded form for billing purposes by the local telephone company pertaining to information on service orders and service equipment records.

Uninterrupted Power Supply (UPS): Designation of a power source that is capable of continuing for a period of time in the absence of utility power.

Utilization Equipment: Equipment that uses electricity.

- V -

Value-Added Network Service (VANS): A data transmission network that routes messages according to available paths, assures that the message will be received as it was sent, and provides for user security, high speed transmission and conferencing among terminals.

Voice Grade (VG): An access line suitable for voice, low-speed data, facsimile, or telegraph service. Generally, it has a frequency range of about 300-3000 Hz.

Volt: The unit of measurement of electrical force. One volt will force one ampere of current to flow through a resistance of one ohm.

Voltage Drop: Voltage reduction due to wire resistance.

- W -

WATT: The unit of measurement of electrical power or rate of work. One watt represents the amount of work that is done by one ampere of current at a pressure of one volt.

Waveform: Characteristic shape of an electrical current or signal.

Wideband: A term applied to facilities or circuits where bandwidths are greater than that required for one voice channel.

Wire Clamp: A device used to secure telephone wires to a surface. One end is U shaped for placement over the wire. The other end contains a tab which is affixed to the mounting surface with a nail or a screw.

Wire Guard: A length of plastic (round or U shaped) use to protect telephone wiring from abrasion or foreign voltages.